Survey Results of the American Public's Values, Objectives, Beliefs, and Attitudes Regarding Forests and Grasslands

DEBORAH J. SHIELDS, INGRED M. MARTIN, WADE E. MARTIN, MICHELLE A. HAEFELE

A Technical Document Supporting the 2000
USDA Forest Service RPA Assessment

Shields, Deborah J.; Martin, Ingrid M.; Martin, Wade E.; and Haefele, Michelle A. 2002. **Survey results of the American public's values, objectives, beliefs, and attitudes regarding forests and grasslands: A technical document supporting the 2000 USDA Forest Service RPA Assessment.** Gen. Tech. Rep. RMRS-GTR-95. Fort Collins, CO: U.S. Department of Agriculture, Forest Service, Rocky Mountain Research Station. 111 p.

Keywords: public lands, public involvement process, values, objectives, beliefs, attitudes, forest and grassland management, stakeholder engagement, strategic planning

Abstract

The USDA Forest Service completed its Strategic Plan (2000 Revision) in October 2000. The goals and objectives included in the Plan were developed with input from the public, some of which was obtained through a telephone survey. We report results of the survey. Members of the American public were asked about their values with respect to public lands, objectives for the management of public lands, beliefs about the role the agency should play in fulfilling those objectives, and attitudes about the job the agency has been doing. The public sees the promotion of ecosystem health as an important objective and role for the agency. There is strong support for protecting watersheds. The public supports multiple uses, but not all uses equally. Motorized recreation is not a high priority objective, while preserving the ability to have a "wilderness experience" is important. There is moderate support for providing resources to dependent communities. The provision of less consumptive services is more important than those that are more consumptive. There is a lack of support for subsidies for development and leasing of public lands. Preservation of traditional uses is a somewhat important objective. Development and use of the best scientific information enjoys wide support, as does information sharing and collaboration. A national direction for the management of National Forest lands is a slightly important objective. Increasing law enforcement on National Forests and Grasslands is an important objective and an appropriate role for the agency. The public has a strong environmental protection orientation, has a moderately strong conservation/preservation orientation, and supports some development.

Cover photos by Lane Eskew, G. Donald Bain, Chistie Van Cleve, and USDA Forest Service, RMRS.

Contents

List of Tables

Executive Summary

Background

The Government Performance and Results Act (Public Law 103-63) requires that each Federal agency submit to Congress a five-year Strategic Plan. The Plan is to include long-term goals and objectives. Identifying the long-term goals and objectives is one of the most critical aspects of Strategic Planning. The Results Act requires an agency to ask for the views and suggestions of anyone "potentially affected by or interested in" its Strategic Plan. The long-term goals and objectives of the Strategic Plan must therefore reflect not only the agency's mission, but also the public's views and suggestions for our country's forests and grasslands. The USDA Forest Service Strategic Plan 2000 Revisions can be found at:

http://www2.srs.fs.fed.us/strategicplan/toc_view_plan.asp

The USDA Forest Service completed its Strategic Plan (2000 Revision) in October 2000. The goals and objectives included in the Plan were developed by the agency with input from the public. This input was obtained in several ways, one of which was through a telephone based survey. This report presents the results of the survey in which approximately 7000 randomly selected members of the American public were asked about their *values* with respect to public lands, *objectives* for the management of forests and grasslands, *beliefs* about the role the USDA Forest Service should play in fulfilling those *objectives*, and *attitudes* about the job the USDA Forest Service has been doing in fulfilling their *objectives*.

Survey results will help the agency understand the public's *objectives*, as well as the underlying value sets that are the basis for these *objectives*. The data on *beliefs* provides information on the degree of importance that the public associates with their objectives. The attitudinal measures provide a useful insight into the public's evaluation of how the USDA Forest Service is meeting or fulfilling these objectives.

Data and Methods

The items in the survey have been extensively pre-tested and applied in various other studies. The *values* scale was designed to focus on values that people hold for public lands (called the Public Lands Values). It was tested using both students and adults around the United States. The *objectives* scale items were developed using input from 80 focus groups around the country. The *beliefs* and *attitudes* scales tier down from the *objectives* items.

The questions in the survey (hereafter referred to as the VOBA survey) are a set of scale items. These are statements to which people are asked to respond using a five-point scale. The *objectives* scale is anchored by 1=not at all important to 5=very important. *Beliefs* are anchored by 1=strongly disagree to 5=strongly agree and *attitudes* are anchored by 1=very unfavorable to 5=very favorable. Each of these three scales consists of 30 items The 25 items in the *values* scale were anchored by 1=strongly disagree to 5=strongly agree.

The objectives statements in the VOBA survey reflect the objectives espoused by the members of the focus groups. All stakeholder interests were represented in the 80 focus groups conducted around the continental United States. Several of the focus group participants had goals that, while not phrased exactly as in the Strategic Plan, were similar in content. It should be noted that these public objectives were arrived at independently of the Strategic Plan Objectives. The results of the broader public survey can then be used to gauge public support for these focus group objectives.

Findings

Demographics

The average age of respondents was 44 years, with approximately 24% aged 30 and under, 41% aged 31 to 50, and 35% over 50. Forty three percent of respondents were male, 57% female. Educational levels were varied, with 11% having less than a high school diploma, 27% having completed high school, 24% having some post-high school education, 26% holding Bachelor's degrees, and 12% having a Master's degree or higher. Average annual household income was $59,000. The NSRE sample over represents non-metropolitan residents with 38% compared to 20% in the general population.

VOBA Results

The results are summarized in three ways. First, the responses are grouped according to their relationship to the USDA Forest Service Strategic Plan. Second, results are organized according to the public's strategic level objectives. Finally, the Public Lands Values are discussed.

Strategic Plan Objectives for the USDA Forest Service

We have identified VOBA objectives that will provide information about the level of support for Strategic Plan Goals and Objectives. In this summary, we speak to the Strategic Plan Goals and highlight selected VOBA objectives that indicate particularly strong support for or disagreement with individual Strategic Plan Goals. Individual Strategic Plan Objectives are addressed in detail in the full report.

Goal 1: Ecosystem Health. There is wide support for the first goal described in the USDA Forest Service Strategic Plan, as the public sees the promotion of ecosystem health as an important objective for public lands and such protection as an important role for the USDA Forest Service.

Goal 2: Multiple Benefits to People. Mixed results are found for the goal of providing multiple benefits to people. The public supports multiple uses but does not support all uses equally. For example, scores for VOBA objectives on motorized recreation indicate that this is not a high priority objective. Conversely, preserving the ability to have a "wilderness experience" is seen asimportant. Finally it should be noted that there is only moderate support for the provision of resources to dependent communities and traditional cultural uses.

Goal 3: Scientific and Technical Assistance. The goal of developing and using the best scientific information enjoys wide support among the public. The public also supports information sharing and collaboration.

Goal 4: Effective Public Service. The delivery of public service elicits a response that is mixed in a similar manner to that of providing benefits to the public. People see the provision of less consumptive services as more important than those that are more consumptive. Especially noteworthy is the lack of support for continued subsidies for development and leasing of public lands.

Strategic Level Objectives of the Public

This section groups the VOBA survey results so as to examine public opinions grouped according to the strategic level objectives expressed by focus group participants during survey development.

Access. The American public is divided in its opinion about the provision of access. This is evidenced by the difference between support for motorized access and support for non-motorized access. The expansion of off-highway motorized access and the development of new paved roads are somewhat unimportant objectives and the provisions of trails for motorized access are slightly unimportant to the public. Contrast this with the provision of non-motorized access which is viewed as a somewhat important objective to the public.

Preservation/Conservation. Protection of ecosystems is seen as an important objective and as an important role for the USDA Forest Service. Especially noteworthy is the strong support for conserving and protecting watersheds, both as a public objective and as a role for the agency.

Economic Development. These items address determination of the level of extractive uses of public lands. These objectives are supported at least moderately by the public.

Education. Providing information to the public about the use, management and conservation of forests and grasslands is considered an important objective for public land management agencies.

Natural Resource Management. Opportunities for public input into the management of forests and grasslands are seen as important by survey respondents. Increased law enforcement, increased acreage in the National Forest system, and diverse uses are also viewed as important objectives by the public.

Public Lands Values

The values scale was examined separately, with the items in the scale grouped into two values dimensions: socially responsible individual values and socially responsible management values. The means for the individual values indicate that the public has a strong orientation toward environmental protection. The results for the management values indicate that the public holds a moderately strong conservation/preservation orientation but that it also supports some degree of development.

Introduction

The USDA Forest Service completed its Strategic Plan (2000 Revision) in October 2000. The Government Performance and Results Act (Public Law 103-62) requires that each agency's Plan include long-term goals and objectives. Identifying the long-term goals and objectives is one of the most critical aspects of Strategic Planning. These objectives must be consistent with the mission of the agency, which is to sustain the health, diversity, and productivity of the Nation's forests and grasslands to meet the needs of present and future generations. To fulfill this mission, the agency not only manages public lands, but assists state and private landowners in the practice of good land stewardship and collaborates with partners and the public as stewards of the forests and grasslands that it holds in trust for the American people. In addition, the USDA Forest Service conducts scientific research on a wide range of subjects related to the performance of the mission described above.

The Results Act requires an agency to ask for the views and suggestions of anyone "potentially affected by or interested in" its Strategic Plan. The long-term goals and objectives of the Strategic Plan must therefore reflect not only the agency's mission, but also the public's views and beliefs for our country's forests and grasslands. This report presents the results of a telephone survey in which randomly selected members of the American public were asked about their:

- *values* with respect to public lands,

- *objectives* for the management, use and conservation of forests and grasslands,

- *beliefs* about the role the USDA Forest Service should play in fulfilling those objectives, and

- *attitudes* about the job the USDA Forest Service has been doing in fulfilling their objectives.

Survey results will help the agency understand the public's *objectives*, as well as the underlying values that are the basis for those *objectives*. The data on beliefs and attitudes will provide information on the importance that various public segments place on the agency's many current and potential activities. The survey on values, objectives, beliefs, and attitudes (VOBA) was implemented as a module of the National Survey on Recreation and the Environment (NSRE), which is conducted by USDA Forest Service[1].

The report of the survey is divided into two sections. The first section will give a brief overview of the survey design, survey implementation, and data analysis methodology. Results of the data analysis are presented in the second section. The results section is further divided into five subsections. The first subsection presents the basic demographic description of the survey results, the second subsection shows the survey results linked to the USDA Forest Service Strategic Plan (2000 Revisions), the third subsection presents the survey results for the values questions, the fourth subsection presents the survey results grouped according to strategic level objectives as expressed by the focus groups during survey development, and the fifth subsection presents the results of a subset of the data broken down according to familiarity with the USDA Forest Service. Methodology is covered in more detail in Appendix A. The values, objectives, beliefs and attitudes survey items are reported in Appendices B through E respectively. Appendix F contains the Forest Service Familiarity Questions, while Appendices G, H and I contain the full results for each subsection in tabular form. A glossary of terms appears in Appendix J.

---

[1] Information on the NSRE is available from Dr. Ken Cordell, USDA Forest Service, Southern Research Station, (706) 559-4263, email kcordell@fs.fed.us.

Methodology

Survey Design and Implementation

Between September 1999 and June 2000, over 80 focus groups and individual interviews were conducted across the lower 48 states. These interviews concentrated on 3 topics: (1) issues related to the use of public lands in general and forests and grasslands in particular, (2) the objectives (or goals) of the group (or individual) regarding the use, management, and conservation of the forests and grasslands, and (3) the role of the Forest Service in the use, management and conservation of the forests and grasslands.

Based upon the results of the focus group interviews, an objectives hierarchy was constructed for each group. These hierarchies indicated what each group or individual was attempting to achieve and how they would achieve each goal or objective. These objectives ranged from the very abstract strategic level to the more focused or concrete means level. The means level objectives are at the bottom of the hierarchy while the strategic objective is at the top. Fundamental objectives between the means level and the strategic level completed the hierarchies. Therefore, the strategic level objective is an abstract objective that can be achieved by more specific fundamental level objectives, which are in turn achieved by means level objectives (see figure 1).

Each of the objectives hierarchies was confirmed with its respective group so as to ensure that it accurately reflected their goals and objectives. A combined objectives hierarchy was then constructed that included all the objectives stated by each group or individual interviewed. The result was a hierarchy that covered 5 strategic level objectives related to access, preservation/conservation, commodity development, education, and natural resource management. These 5 strategic level objectives were supported by 30 fundamental objectives.

The 30 fundamental level objectives were used to develop 30 objectives statements that were used in the National Survey of Recreation and the Environment (NSRE). The NSRE is a national survey administered via telephone interviews. The 30 objectives statements were divided into 5 groups based upon the strategic level objectives that the focus groups had identified. During the telephone interviews, each respondent was asked one statement from

Figure 1. Objectives hierarchy.

each of the 5 strategic level groups in order to obtain a statistically valid sample for each statement and for each strategic level group.

As noted above, the survey of the American public's values, objectives, beliefs, and attitudes (hereafter VOBA) was conducted as a module within the NSRE. Questions about respondents' recreation behavior comprise the bulk of the interview; however, the results presented here are based solely on the questions in the VOBA Module of the survey and the demographic questions. The VOBA questions are sets of scale items which people are asked to respond using a five-point scale. The objectives items are anchored by 1=not at all important to 5=very important. Beliefs are anchored by 1=strongly disagree to 5=strongly agree and attitudes are anchored by 1=very unfavorable to 5=very favorable. Each of these three types of scales consist of 30 items. The 25 items in the values scale are anchored by 1=strongly disagree and 5=strongly agree. The full four-part survey, comprised of 115 items, can be found in Appendices B-E. For detail on methods see Appendix A.[2]

Data Analysis

This report is based on 7,069 responses to the NSRE phone survey. The data do not accurately reflect the demographics of the United States and so were weighted prior to analysis to account for this. An overall weight was constructed based on the following demographic factors: (1) age, sex, and race category (60 cells total), (2) education level, and (3) metropolitan or non-metropolitan area. These weights were constructed using the same method developed for use with the full NSRE data set. Weighted means and standard deviations were then calculated for each of the 115 items.[3]

In order to characterize the strength of preferences, modifiers have been assigned to ranges for the item means and group means. An objective is characterized as "very important" if the mean is 4.75 or above. Respondents are said to "strongly agree" that an item is an appropriate role for the USDA Forest Service (or that the item is a "very important" role) if the mean falls into this range. Attitudes are "very favorable" if the mean is 4.75 or higher. No modifier is attached to means between 4.00 and 4.75 (objectives are "important," respondents "agree," or the role is "important," and attitudes are "favorable"). If the mean falls into the range 3.25 to 4.00 objectives are "somewhat important," respondents "agree somewhat" that the item is a "somewhat important role" for the agency and attitudes are "somewhat favorable." "Slightly" is used to describe objectives, attitudes and beliefs which fall between 3.00 and 3.25. A mean of 3.00 is neutral. Objectives that fall into the range between 2.75 and 3.00 are "slightly unimportant," a mean for beliefs in this range implies that respondents "disagree slightly" (or the item is a "slightly unimportant" role for the agency) and attitudes in this range are "slightly unfavorable." No modifier will be attached to objectives, beliefs, or attitudes between 1.25 and 2.00. Attitudes 1.25 or below are "very unfavorable." Responses to beliefs items below 1.25 indicate that the public "strongly disagrees" that the item is an appropriate role (or the role is "very unimportant") and objectives in this range are "very unimportant" (see table 1).

The items in the national survey instrument have been grouped into various subsets to facilitate interpretation and analysis. These groups will be presented in three subsections

---

[2] For a more in-depth discussion of how the scale items were developed and tested see Martin, Martin, Shields and Wise (1999) and Martin, Martin, and Shields (2000).

[3] The weighting method was originally developed for use with the full NSRE data set. Additional information on the weighting methodology is available from Carter Betz, USDA Forest Service, Southern Research Station, email cbetz@fs.fed.us or Dr. Michelle Haefele, Department of Economics, Colorado State University, email mhaefele@fs.fed.us.

Table 1. Modifiers used to describe survey response scores.

Range of mean	Objectives for the management of forests and grasslands	Beliefs about the role the USDA Forest Service should have in fulfilling objectives	Attitudes about the performance of the USDA Forest Service in fulfilling objectives
1.00 – 1.25	Objective is very unimportant	Role is very unimportant (or very inappropriate)	Performance is very unfavorable
1.25 – 2.00	No modifier (unimportant)	No modifier (role is unimportant or inappropriate)	No modifier (performance is unfavorable)
2.00 – 2.75	Objective is somewhat unimportant	Role is somewhat unimportant (or somewhat inappropriate)	Performance is somewhat unfavorable
2.75 – 3.00	Objective is slightly unimportant	Role is slightly unimportant (or slightly inappropriate)	Performance is slightly unfavorable
3.00 – 3.25	Objective is slightly important	Role is slightly important (or slightly appropriate)	Performance is slightly favorable
3.25 – 4.00	Objective is somewhat important	Role is somewhat important (or somewhat appropriate)	Performance is somewhat favorable
4.00 – 4.75	No modifier (objective is important)	No modifier (role is important or appropriate)	Non modifier (performance is favorable)
4.75 – 5.00	Objective is very important	Role is very important (or very appropriate)	Performance is very favorable

within the Results section of this report. The three sections are: Survey Results Linked to USDA Forest Service Strategic Plan Goals and Objectives, Public Lands Values (PLV), and Survey Objectives, Beliefs, and Attitudes Grouped by Strategic Level Objectives.

For all of the groupings of the data, responses to the demographic questions were used to break the full data set down, first by region (east and west, divided at approximately the 100th meridian), then by metropolitan and non-metropolitan counties (based upon Census classification) within each region.

The first 5,064 respondents were asked five questions designed to determine their familiarity with the USDA Forest Service. Responses to these questions were used to construct an overall measure of familiarity. Respondents who correctly answered four or more of these questions were designated as having a high degree of familiarity with the USDA Forest Service. Those answering two or three correctly were designated as having a moderate level of familiarity. Respondents who answered fewer than two of the questions correctly were designated as having low agency familiarity. Results for this subset of respondents are broken down by degree of familiarity and analyzed in a separate subsection of the Results section. These are grouped as described above.

The weighted means for each item, divided by the subsets described above, and by the demographic breakdowns, are presented in Appendices G (by Strategic Plan objectives), H (Public Lands Values), and I (by Strategic Level Objectives).

Results

A. Demographic Description of Sample

The objective of the first step in the analysis was to develop an overall picture of the NSRE respondents through examination of the demographic data. The average age of respondents is 44 years, with approximately 24% aged 30 and under, 41% aged 31 to 50, and 35% aged 51 and older. Forty-three percent (3,019) of respondents were male, 57% (4,038) female. Educational level was varied with 753 (11%) having less than a high school diploma, 1,873 (27%) having completed high school, 1,706 (24%) having some post-high school education, 1,816 (26%) holding bachelor's degrees, and 828 (12%) having a master's degree or higher. Reporting income is less straight forward because some respondents gave actual income figures while others would only identify their income category. Based on both types of responses, average annual household income for all respondents was estimated at $59,000/year. A comparison between the NSRE sample demographics and the most recent census estimates is presented in table 2.

Table 2. Comparison of National Survey on Recreation and the Environment demographics with Census demographics.

		NSRE Sample	Census
Age:	30 and under	24%	43%
	31-50	41%	31%
	51 and older	35%	27%
Sex:	male	43%	49%
	female	57%	51%
Education:	less than high school	11%	17%
	high school diploma	27%	31%
	some college	24%	18%
	bachelor s degree	26%	24%
	post-grad degree	12%	7%
Residence:	metro counties	62%	80%
	non-metro counties	38%	20%

B. Survey Results Linked to USDA Forest Service Strategic Plan

This section presents selected survey results linked to the USDA Forest Service Strategic Plan goals and objectives. The goal of the national VOBA survey was to provide input for revising the National Strategic Plan for the USDA Forest Service. The data collected regarding the objectives of the American public during the first stage of surveying were matched with the existing Strategic Plan to provide additional input to the agency regarding the existing plan. The text which follows presents only the results for the full sample. Breakdowns by demographic characteristics for the VOBA items by Strategic Plan Objectives can be found in Appendix G.

Goal 1: Ecosystem Health: Promote ecosystem health and conservation using a collaborative approach to sustain the Nation's forests, grasslands and watersheds.

The USDA Forest Service will promote ecosystem health and conservation using a collaborative approach to sustain the Nation's forests, grasslands, and watersheds. Supporting this goal are three specific objectives.

Strategic Plan Objective 1.a: Improve and protect watershed conditions to provide the water quality and quantity and the soil productivity necessary to support ecological functions and intended beneficial water use.

The first objective supporting ecosystem health deals with protecting water and soil resources. The public responses for the items in the VOBA that support this objective are found in table 3.

The public sees the protection of watersheds as an important objective for public land management. In the West, those in metropolitan areas see this as a more important objective than do their non-metropolitan counterparts. The public also agrees that this is an important role for the USDA Forest Service, with metropolitan dwellers in the West viewing this as slightly more important role for the agency than those in the non-metropolitan West. The agency is also given a favorable evaluation for its performance in this area, with metropolitan westerners giving higher ratings than non-metropolitan westerners.

The public sees the development of volunteer programs that improve the quality of public lands as an important objective and people see this as an important role for the agency. Residents of metropolitan areas in both the East and West see this as a slightly more important objective than do those in non-metropolitan areas. The agency is given a somewhat favorable performance rating for facilitating such volunteer programs.

Table 3. National mean scores for Strategic Plan Objective 1.a.: Improve and protect watershed conditions to provide the water quality and quantity and the soil productivity necessary to support ecological functions and intended beneficial water uses. Survey items are from the VOBA module of the National Survey on Recreation and the Environment.

Survey item number and statement	Objectives for the management of forests and grasslands (1=not at all important, 5=very important)	Beliefs about the role of the USDA Forest Service (1=strongly disagree, 5=strongly agree)	Attitudes about the performance of the Forest Service (1=very unfavorable, 5=very favorable)
7. Conserve and protect watersheds	4.73 *0.76*[a]	4.61 *0.83*	3.91 *1.17*
18. Develop volunteer programs to improve land (tree planting, etc.)	4.60 *0.86*	4.46 *1.03*	3.85 *1.25*

[a] Standard deviation

Strategic Plan Objective 1.b: Provide ecological conditions to sustain viable populations of native and desired nonnative species and to achieve objectives for Management Indicator Species (MIS)/focal species.

Habitat protection is the second objective under the goal of ecosystem health. The responses to these items can be found in table 4.

Overall, the protection of ecosystems and wildlife habitat is seen as an important objective for public land management. This objective is more important to those respondents living in metropolitan areas than to those in non-metropolitan areas. This difference is especially pronounced in the West, with the exception of restricting mineral development. Restriction of mineral development is a more important objective for metropolitan easterners than it is for metropolitan westerners. Respondents also believe that habitat protection is an important role for the USDA Forest Service, and again metropolitan dwellers (both East and West) see this as a more important role for the agency. The USDA Forest Service is doing a somewhat favorable job in fulfilling this objective.

Some of the means by which ecosystem health and habitat protection are to be achieved (wilderness designation and the restriction of various extractive resource uses) are seen as at least somewhat important objectives by the public and as somewhat important roles for the USDA Forest Service. In cases where a significant difference between demographic groups appears, metropolitan residents and Easterners are more supportive of such restrictive policies.

Table 4. National mean scores for Strategic Plan Objective 1.b.: Provide ecological conditions to sustain viable populations of native and desired nonnative species and to achieve objectives for Management Indicator Species (MIS)/focal species. Survey items are from the VOBA module of the National Survey on Recreation and the Environment.

Survey item number and statement	Objectives for the management of forests and grasslands (1=not at all important, 5=very important)	Beliefs about the role of the USDA Forest Service (1=strongly disagree, 5=strongly agree)	Attitudes about the performance of the Forest Service (1=very unfavorable, 5=very favorable)
6. Designate more wilderness to stop development & motorized access	3.84 _1.41_[a]	3.66 _1.46_	3.45 _1.34_
8. Preserve natural resources through policies such as no timber, no mining	4.22 _1.23_	4.21 _1.27_	3.65 _1.31_
13. Restrict development of minerals	3.96 _1.42_	3.95 _1.43_	3.30 _1.50_
14. Restrict timber harvest and grazing	3.99 _1.27_	3.94 _1.34_	3.50 _1.36_

[a] Standard deviation

Strategic Plan Objective 1.c: Increase the amount of forests and grasslands restored to or maintained in a healthy condition with reduced risk and damage from fires, insects and diseases, and invasive species.

The restoration and maintenance of ecosystem health achieved through protection and volunteer programs is an important objective for the public. See table 5.

Table 5. National mean scores for. Strategic Plan Objective 1.c.: Increase the amount of forests and grasslands restored to or maintained in a healthy condition with reduced risk and damage from fires, insects and diseases, and invasive species. Survey items are from the VOBA module of the National Survey on Recreation and the Environment.

Survey item number and statement	Objectives for the management of forests and grasslands (1=not at all important, 5=very important)	Beliefs about the role of the USDA Forest Service (1=strongly disagree, 5=strongly agree)	Attitudes about the performance of the Forest Service (1=very unfavorable, 5=very favorable)
9. Protecting ecosystems and wildlife habitat	4.58 _0.92_[a]	4.53 _0.98_	3.90 _1.16_
18. Develop volunteer programs to improve land (tree planting, etc.)	4.60 _0.86_	4.46 _1.03_	3.85 _1.25_

[a] Standard deviation

Metropolitan residents in both the East and West see the objective of protecting ecosystems and wildlife habitat as more important than do those in non-metropolitan areas. Within non-metropolitan areas, those in the East are more in favor of such programs than are westerners.

The public sees these activities as important roles for the USDA Forest Service and views the agency performance as somewhat favorable.

Goal 2: Multiple Benefits to People: Provide a variety of uses, values, products, and services for present and future generations by managing within the capability of sustainable ecosystems.

The second broad goal outlined in the Strategic Plan deals with the provision of benefits to the American public. The USDA Forest Service seeks to provide a variety of uses, values, products, and services for present and future generations by managing within the capability of sustainable ecosystems. This goal is supported by five specific objectives.

Strategic Plan Objective 2.a: Improve the capability of the Nation's forests and grasslands to provide diverse, high-quality outdoor recreation opportunities.

The first Strategic Plan objective deals with providing outdoor recreation opportunities to the public. The results for the VOBA questions dealing with this recreation objective are in table 6.

Recreation opportunities can be divided into motorized and non-motorized. The provision of increased access for motorized recreation is seen as a slightly unimportant objective for public land management and is also viewed as a slightly unimportant role for the USDA

Table 6. National mean scores for Strategic Plan Objective 2.a.: Improve the capability of the Nation's forests and grasslands to provide diverse, high-quality outdoor recreation opportunities. Survey items are from the VOBA module of the National Survey on Recreation and the Environment.

Survey item number and statement	**Values** with respect to forests and grasslands (1=strongly disagree 5=strongly agree)	**Objectives** for the management of forests and grasslands (1=not at all important 5= very important)	**Beliefs** about the role of the USDA Forest Service (1=strongly disagree 5=strongly agree)	**Attitudes** about the performance of the Forest Service (1=very unfavorable 5=very favorable)
1. Expand off-highway motorized access		2.41 *1.49[a]*	2.52 *1.50*	2.97 *1.45*
2. Increase trails for motorized vehicles		2.82 *1.47*	2.98 *1.51*	3.25 *1.38*
5. Develop new paved roads		2.62 *1.49*	2.70 *1.57*	3.19 *1.43*
3. Increase trails for non-motorized recreation		3.75 *1.33*	3.71 *1.37*	3.59 *1.27*
4. Designate trails for specific use		3.59 *1.43*	3.94 *1.25*	3.61 *1.33*
19. Develop volunteer programs to improve facilities (trails, etc.)		4.18 *1.13*	4.22 *1.09*	3.79 *1.21*
20. Inform public about recreation concerns (safety, trail etiquette, etc.)		4.53 *0.93*	4.50 *0.93*	3.89 *1.31*
28. Collect an entry fee to support National Forests and Grasslands		3.66 *1.36*	3.69 *1.43*	3.61 *1.36*
9. I would pay $5 more to use public lands for recreation	3.49 *1.60*			

[a] Standard deviation

Forest Service. Non-metropolitan easterners and metropolitan westerners see motorized access as a more important objective than do non-metropolitan westerners and metropolitan easterners. Agency performance in the area of motorized recreation is viewed as somewhat favorable, except in the case of off-highway motorized access, where the agency role and performance are rated as slightly unfavorable.

The provision of opportunities and facilities for non-motorized recreation is seen as a somewhat important objective and as a somewhat important role for the agency. The agency is viewed as doing a somewhat favorable job providing non-motorized access. Support for non-motorized recreation opportunities is stronger in metropolitan areas than non-metropolitan areas. Separating these often conflicting types of pursuits by designating trails for specific uses is seen as a somewhat important objective, with higher support outside of metropolitan areas.

The use of volunteers to improve recreation facilities such as campgrounds, picnic areas and trails is an important objective, as well as an important role for the USDA Forest Service. The strongest support for such programs appears in the western portion of the country. Survey respondents also feel that it is important for the agency to provide information to the public about recreation concerns such as trail etiquette and safety.

The expansion of commercial recreation is seen as a slightly important objective and a slightly important role for the agency. The agency is providing a somewhat favorable level of commercial recreation.

The collection of recreation fees is seen as a somewhat important objective and a somewhat important agency role. Responses to a values statement regarding a payment of $5 extra to use public lands for recreation indicates some agreement, with stronger agreement in both eastern and western metropolitan areas than in the non-metropolitan areas.

Strategic Plan Objective 2.b: Improve the capability of wilderness and protected areas to sustain a desired range of benefits and values.

The second objective in providing benefits to people is to provide a desired range of values and benefits from wilderness areas. The VOBA results under this objective are in table 7.

Generally, the preservation of a "wilderness" experience and the designation of wilderness are seen by the public as at least somewhat important objectives to pursue. Metropolitan residents show the strongest support for this objective in both the East and West. Among

Table 7. National mean scores for Strategic Plan Objective 2.b.: Improve the capability of wilderness and protected areas to sustain a desired range of benefits and values. Survey items are from the VOBA module of the National Survey on Recreation and the Environment.

Survey item number and statement	Objectives for the management of forests and grasslands (1=not at all important, 5=very important)	Beliefs about the role of the USDA Forest Service (1=strongly disagree, 5=strongly agree)	Attitudes about the performance of the Forest Service (1=very unfavorable, 5=very favorable)
6. Designate more wilderness to stop development & motorized access	3.84 *1.41*[a]	3.66 *1.46*	3.45 *1.34*
10. Preserve "wilderness" experience	4.15 *1.28*	4.22 *1.14*	3.88 *1.10*
9. Protecting ecosystems and wildlife habitat	4.58 *0.92*	4.53 *0.98*	3.90 *1.16*
8. Preserve natural resources through policies such as no timber, no mining	4.22 *1.23*	4.21 *1.27*	3.65 *1.31*

[a] Standard deviation

non-metropolitan residents, those in the East show stronger support than those in the West for the designation of more wilderness. These activities are at least somewhat important roles for the USDA Forest Service, and the agency is doing a somewhat acceptable job.

The protection of ecosystems and habitats and the restriction of natural resource extraction are also seen as important objectives. The public also sees these activities as appropriate roles for the USDA Forest Service. Again, where statistically significant differences between demographic groups occur, metropolitan dwellers are more in favor of such policies than are those living in non-metropolitan areas. The agency is doing a somewhat adequate job of protecting ecosystems and restricting extractive uses.

Strategic Plan Objective 2.c: Improve the capability of the Nation's forests and grasslands to provide desired sustainable levels of uses, values, products, and services.

The third objective supporting the goal of providing benefits to people deals with the provision of products and services. These results can be found in table 8.

The American public sees the provision of natural resources to communities dependent on public land outputs as a somewhat important objective. It is not surprising that those in the non-metropolitan West see this objective as more important than do those in the metropolitan West. The provision of resources is also a somewhat important role for the USDA Forest Service. Making the permitting process easier for established users could facilitate the provision of resources to dependent communities, but the public sees this as a slightly unimportant objective. Those in the non-metropolitan West are also more supportive of relaxing the permitting process. The agency is doing a somewhat favorable job of fulfilling these objectives.

The preservation of local cultural uses of public lands is a somewhat important objective for respondents and is believed to be a somewhat important role for the agency. The need to allow for diverse uses of public lands is an important objective as well as an important USDA Forest Service role.

Table 8. National mean scores for Strategic Plan Objective 2.c.: Improve the capability of the Nation's forests and grasslands to provide desired sustainable levels of uses, values, products, and services. Survey items are from the VOBA module of the National Survey on Recreation and the Environment.

Survey item number and statement	Objectives for the management of forests and grasslands (1=not at all important, 5=very important)	Beliefs about the role of the USDA Forest Service (1=strongly disagree, 5=strongly agree)	Attitudes about the performance of the Forest Service (1=very unfavorable, 5=very favorable)
12. Provide natural resources to dependent communities	3.60 *1.39*[a]	3.29 *1.39*	3.45 *1.30*
15. Make the permitting process easier for established uses	2.89 *1.55*	2.90 *1.59*	3.05 *1.43*
11. Preserve cultural uses	3.82 *1.38*	3.75 *1.36*	3.39 *1.37*
25. Allow for diverse uses	4.07 *1.18*	4.02 *1.14*	3.73 *1.18*
18. Develop volunteer programs to improve land (tree planting, etc.)	4.60 *0.86*	4.46 *1.03*	3.85 *1.25*
16. Develop a national policy for natural resource development	4.26 *1.23*	4.21 *1.19*	3.51 *1.26*
27. Increase total number of acres in the National Forest and Grassland system	3.81 *1.42*	3.95 *1.47*	3.52 *1.45*

[a] Standard deviation

While respondents feel that it is an important objective to keep management decisions local (see table 9), they also feel that it is important to develop a national policy for natural resource development. Metropolitan dwellers see the development of a national natural resource policy as a more important objective than do non-metropolitan residents. The development of such a policy is viewed as an important role for the USDA Forest Service, which is seen as doing a somewhat favorable job.

An indirect objective toward improving the provision of goods and services is increasing the total number of acres in the National Forest System. This objective is seen as somewhat important by survey respondents, and it is seen as a somewhat important role for the USDA Forest Service. Easterners and metropolitan residents are more supportive of increasing NFS acreage. The agency performance is viewed as somewhat favorable.

Table 9. National mean scores for Strategic Plan Objective 2.d.: Increase accessibility to a diversity of people and members of underserved and low-income populations to the full range of uses, values, products, and services. Survey items are from the VOBA module of the National Survey on Recreation and the Environment.

Survey item number and statement	Objectives for the management of forests and grasslands (1=not at all important, 5=very important)	Beliefs about the role of the USDA Forest Service (1=strongly disagree, 5=strongly agree)	Attitudes about the performance of the Forest Service (1=very unfavorable, 5=very favorable)
11. Preserve cultural uses	3.82 *1.38*[a]	3.75 *1.36*	3.39 *1.37*
12. Provide natural resources to dependent communities	3.60 *1.39*	3.29 *1.39*	3.45 *1.30*
15. Make the permitting process easier for established uses	2.89 *1.55*	2.90 *1.59*	3.05 *1.43*
16. Develop a national policy for natural resource development	4.26 *1.23*	4.21 *1.19*	3.51 *1.26*
26. Make management decisions at local level (rather than national)	3.93 *1.22*	3.88 *1.34*	3.49 *1.32*

[a] Standard deviation

Strategic Plan Objective 2.d: Increase accessibility to a diversity of people and members of underserved and low-income populations to the full range of uses, values, products, and services.

Increasing accessibility for populations traditionally underserved by the USDA Forest Service is the fourth objective under the goal of providing benefits to people. The VOBA results are presented in table 9.

The preservation of local cultural uses is a somewhat important objective for the public and a somewhat important role for the agency. Again, while the public sees the provision of natural resources to dependent communities as a somewhat important objective for public lands management, they do not see making the permitting process easier as an important means to achieve this. Easterners, both non-metropolitan and metropolitan, give the agency a higher rating on providing resources for dependent communities. Metropolitan dwellers in the East view this as a more important objective than do those in the metropolitan West.

Another interesting finding emerges when examining the level of importance placed on the objective of making decisions locally rather than nationally and the development of a national natural resource policy (both are considered at least somewhat important by the public). Metropolitan residents in both the East and West feel that making a national policy for resource development is a more important objective and role for the agency than do non-metropolitan residents. Among non-metropolitan residents, those in the East are more in favor of a national policy than those in the West.

Strategic Plan Objective 2.e: Improve delivery of services to urban communities.

The final objective supporting the goal of providing benefits to people is the improvement of delivery of services to urban communities. Table 10 shows the results for the sample as a whole and broken down according to metropolitan and non-metropolitan counties.

Services to metropolitan areas include the provision of recreational opportunities, which can be divided into motorized and non-motorized forms. Increasing motorized opportunities can be achieved through the expansion of motorized access and paved roads. All groups see the expansion of off-highway motorized access as a somewhat unimportant objective and they all believe it is a somewhat unimportant role for the USDA Forest Service. The agency is seen as doing a slightly unfavorable job at the national level, while non-metropolitan residents rate the performance as slightly favorable. The development of new paved roads is also seen as a somewhat unimportant objective by all groups, with the exception of those in the metropolitan West, who give this objective only a slightly unimportant rating. All groups view the development of new paved roads as a somewhat unimportant role for the agency, with the exception of metropolitan westerners who view this role as slightly important. All groups rate the agency performance as slightly favorable. Providing trails for motorized vehicles is viewed as a slightly unimportant objective at the national level, with non-metropolitan residents (particularly those in the East) viewing it as a slightly important objective. At the national level, trails for motorized vehicles are seen as a slightly unimportant role for the USDA Forest Service. Those in the metropolitan West see this role as more important than their non-metropolitan western counterparts. Agency performance is seen as at least slightly favorable by all groups.

Table 10. National mean scores for Strategic Plan Objective 2.e.: Improve delivery of services to urban communities. Survey items are from the VOBA module of the National Survey on Recreation and the Environment.

Survey item number and statement	Objectives for the management of forests and grasslands (1=not at all important, 5=very important)			Beliefs about the role of the USDA Forest Service (1=strongly disagree, 5=strongly agree)			Attitudes about the performance of the Forest Service (1=very unfavorable, 5=very favorable)		
	Full sample	Non-metro	Metro	Full sample	Non-metro	Metro	Full sample	Non-metro	Metro
1. Expand off-highway motorized access	2.41 *1.49[a]*	2.47 *1.07*	2.40 *1.71*	2.52 *1.50*	2.68 *1.13*	2.49 *1.66*	2.97 *1.45*	3.03 *1.02*	2.95 *1.68*
5. Develop new paved roads	2.62 *1.49*	2.54 *1.11*	2.64 *1.70*	2.70 *1.57*	2.55 *1.03*	2.73 *1.83*	3.19 *1.43*	3.12 *1.00*	3.21 *1.65*
2. Trails for motorized vehicles	2.82 *1.47*	3.02 *1.08*	2.78** *1.65*	2.98 *1.51*	2.92 *1.12*	3.00 *1.72*	3.25 *1.38*	3.24 *0.94*	3.26 *1.58*
3. Trails for non-motorized recreation	3.75 *1.33*	3.48 *0.92*	3.81*** *1.51*	3.71 *1.37*	3.38 *1.05*	3.78*** *1.52*	3.59 *1.27*	3.48 *0.98*	3.61 *1.44*
4. Designate trails for specific use	3.59 *1.43*	3.75 *0.99*	3.56* *1.64*	3.94 *1.25*	4.08 *0.93*	3.91 *1.41*	3.61 *1.33*	3.75 *0.90*	3.57* *1.56*
10. Preserve "wilderness" experience	4.15 *1.28*	4.03 *0.95*	4.18* *1.44*	4.22 *1.14*	4.10 *0.90*	4.24 *1.26*	3.88 *1.10*	3.89 *0.81*	3.88 *1.24*
17. Expand commercial recreation	3.04 *1.45*	3.08 *1.01*	3.04 *1.67*	3.25 *1.53*	3.10 *1.04*	3.28 *1.76*	3.45 *1.24*	3.32 *0.95*	3.48* *1.40*

[a] Standard deviation
*, **, *** Mean difference is significant at α = 0.05, 0.01, 0.001

Providing trails for non-motorized recreation is seen as a somewhat important objective. Metropolitan dwellers see this as more important than do non-metropolitan residents in both the East and West. Among non-metropolitan residents, those in the East are more supportive of trails for non-motorized recreation than those in the West. This is seen as a somewhat important role for the USDA Forest Service, again more so by those in metropolitan areas. Agency performance is rated as somewhat favorable.

Separating often conflicting motorized and non-motorized uses by designating trails for specific purposes is a somewhat important objective, with the most support coming from the non-metropolitan West. This is seen as a somewhat important role for the USDA Forest Service, again, especially by those in the non-metropolitan West. Agency performance is somewhat favorable with the strongest support in the non-metropolitan East.

The preservation of a wilderness experience is an important objective for all groups, especially those living in eastern metropolitan areas. All groups see this as an important role for the agency. An interesting result is the fact that those in the non-metropolitan East see this as a more important role than do those in the non-metropolitan West. The agency performance is rated as somewhat favorable.

The public sees the expansion of commercial recreation as a slightly important objective with the strongest support in the metropolitan West. Commercial recreation is seen as a slightly important role for the agency. Agency performance is rated as somewhat favorable.

Goal 3: Scientific and Technical Assistance: Develop and use the best scientific information available to deliver technical and community assistance and to support ecological, economic, and social sustainability.

This goal states that the USDA Forest Service will develop and use the best scientific information available to deliver technical and community assistance and to support ecological, economic, and social sustainability.

Strategic Plan Objective 3.a: Better assist in building the capacity of Tribal governments, rural communities, and private landowners to adapt to economic, environmental, and social change related to natural resources.

Table 11 shows the VOBA results for the first objective under the goal of providing scientific and technical assistance.

This first Strategic Plan objective deals with increasing the ability of tribal governments, rural communities, and private landowners affected by public land use decisions to adapt to changes related to these decisions. Objectives that were presented to the NSRE respondents included several dealing with the provision of information on the impacts and concerns related to the use of natural resources. Specifically, information on recreation concerns, information on the environmental impacts of uses, and information on the economic value of developing natural resources were all seen as important objectives for public land management. Metropolitan residents in both the East and West are more strongly supportive of providing information about recreation concerns and environmental impacts than are those in non-metropolitan areas. Regional differences among non-metropolitan residents shows that those in the East are also more supportive of providing information about environmental impacts than are westerners. Eastern metropolitan residents see the provision of information on the economic value of resource development as a less important objective than those in the non-metropolitan East. The provision of information is seen as an important role for the USDA Forest Service by all groups. The agency is seen as performing at least somewhat favorably by all groups in the role of information dissemination.

NSRE respondents feel that it is an important objective for public land managers to encourage collaboration between groups, with no difference between non-metropolitan and metropolitan residents. Easterners (both non-metropolitan and metropolitan) are more

Table 11. National mean scores for Strategic Plan Objective 3.a.: Better assist in building the capacity of Tribal governments, rural communities, and private landowners to adapt to economic, environmental, and social change related to natural resources. Survey items are from the VOBA module of the National Survey on Recreation and the Environment.

Survey item number and statement	Objectives for the management of forests and grasslands (1=not at all important, 5=very important)			Beliefs about the role of the USDA Forest Service (1=strongly disagree, 5=strongly agree)			Attitudes about the performance of the Forest Service (1=very unfavorable, 5=very favorable)		
	Full sample	Non-metro	Metro	Full sample	Non-metro	Metro	Full sample	Non-metro	Metro
20. Inform public about recreation concerns (safety, trail etiquette, etc.)	4.53 *0.93[a]*	4.43 *0.79*	4.56[*] *1.00*	4.50 *0.93*	4.47 *0.71*	4.51 *1.05*	3.89 *1.31*	3.95 *0.90*	3.88 *1.50*
21. Inform public on potential environmental impacts of uses	4.39 *1.02*	4.17 *0.87*	4.45[***] *1.10*	4.48 *1.00*	4.36 *0.74*	4.50[*] *1.13*	3.50 *1.33*	3.55 *0.95*	3.49 *1.53*
23. Encourage collaboration between groups to share information	4.15 *1.20*	4.14 *0.82*	4.15 *1.39*	4.15 *1.17*	4.17 *0.83*	4.14 *1.34*	3.72 *1.22*	3.70 *0.88*	3.72 *1.42*
24. Using public advisory committees to advise on management issues	3.90 *1.20*	3.79 *0.90*	3.93[*] *1.36*	3.84 *1.25*	4.02 *0.87*	3.80[*] *1.42*	3.36 *1.26*	3.38 *0.90*	3.35 *1.46*
26. Make management decisions at local level (rather than national)	3.93 *1.22*	4.03 *0.85*	3.90 *1.40*	3.88 *1.34*	4.14 *0.91*	3.81[**] *1.53*	3.49 *1.32*	3.61 *0.97*	3.47 *1.50*
11. Preserve cultural uses	3.82 *1.38*	3.76 *1.00*	3.84 *1.57*	3.75 *1.36*	3.65 *1.06*	3.78 *1.52*	3.39 *1.37*	3.53 *0.96*	3.36[*] *1.58*

[a] Standard deviation

*, **, *** Mean difference is significant at $\alpha = 0.05, 0.01, 0.001$

supportive of this objective than westerners. The public also feels that this is an important role for the USDA Forest Service. Agency performance is rated as somewhat favorable.

Public involvement in management issues through advisory committees is seen as a somewhat important objective, with slightly stronger support among metropolitan residents. The use of advisory committees is seen as a somewhat important role for the USDA Forest Service, which is seen as doing a somewhat favorable job in performing this role.

The public feels that it is at least somewhat important that management decisions be made at the local level. Non-metropolitan residents in the West are most supportive of this as an important role for the agency, with all groups viewing it as at least somewhat important. The USDA Forest Service performance is rated as somewhat favorable.

The preservation of local cultural uses of National Forest lands is seen as a somewhat important objective by all groups. This is also seen as a somewhat important role for the agency. Agency performance is rated as somewhat favorable by all groups.

Strategic Plan Objective 3.b: Increase the effectiveness of scientific, developmental, and technical information delivered to domestic and international interests.

The second Strategic Plan objective supporting the goal of providing scientific and technical assistance concerns the effectiveness of such assistance. See table 12 for VOBA items related to this Strategic Plan objective.

Table 12. National mean scores for Strategic Plan Objective 3.b. Increase the effectiveness of scientific, developmental, and technical information delivered to domestic and international interests. Survey items are from the VOBA module of the National Survey on Recreation and the Environment.

Survey item number and statement	Objectives for the management of forests and grasslands *(1=not at all important, 5=very important)*	Beliefs about the role of the USDA Forest Service *(1=strongly disagree, 5=strongly agree)*	Attitudes about the performance of the Forest Service *(1=very unfavorable, 5=very favorable)*
20. Inform public about recreation concerns (safety, trail etiquette, etc.)	4.53 *0.93*[a]	4.50 *0.93*	3.89 *1.31*
21. Inform public on potential environmental impacts of uses	4.39 *1.02*	4.48 *1.00*	3.50 *1.33*
22. Inform public on economic value from developing natural resources	4.02 *1.30*	4.08 *1.17*	3.40 *1.38*
23. Encourage collaboration between groups to share information	4.15 *1.20*	4.15 *1.17*	3.72 *1.22*

[a] Standard deviation

As seen above, the provision of information on recreation concerns, information on the environmental impacts of uses, and information on the economic value of developing natural resources were all seen as important objectives for public land management. Metropolitan residents in both the East and West are more strongly supportive of providing information about recreation concerns and environmental impacts than are those in non-metropolitan areas. Regional differences among both non-metropolitan and metropolitan residents shows that those in the East are also more supportive of providing information about environmental impacts than are non-metropolitan westerners. Eastern metropolitan residents see the provision of information on the economic value of resource development as more important than do those in the non-metropolitan East. The provision of information is seen as an important role for the USDA Forest Service by all groups. The agency is seen as performing at least somewhat favorably by all groups in the role of information dissemination.

The public also feels that it is an important objective for public land managers to encourage collaboration between groups, with no difference between non-metropolitan and metropolitan residents. Easterners (both non-metropolitan and metropolitan) are more supportive of this objective. The public also feels that this is an important role for the USDA Forest Service.

Strategic Plan Objective 3.c: Improve the knowledge base provided through research, inventory, and monitoring to enhance scientific understanding of ecosystems, including human uses, and to support decision-making and sustainable management of the Nation's forests and grasslands.

The provision of scientific and technical assistance includes research efforts to improve the understanding of ecosystems, as described in the third Strategic Plan objective under this goal. VOBA results related to this objective can be found in table 13.

Providing the public with information about the human, economic, and environmental impacts associated with various uses of National Forest lands are all seen as important objectives by the public. The public agrees that the USDA Forest Service is an appropriate agency to provide this information and they are doing a somewhat adequate job. Similar results are found for the objective of encouraging collaboration between groups.

Table 13. National mean scores for Strategic Plan Objective 3.c.: Improve the knowledge base provided through research, inventory, and monitoring to enhance scientific understanding of ecosystems, including human uses, and to support decision-making and sustainable management of the Nation's forests and grasslands. Survey items are from the VOBA module of the National Survey on Recreation and the Environment.

Survey item number and statement	Objectives for the management of forests and grasslands (1=not at all important, 5=very important)	Beliefs about the role of the USDA Forest Service (1=strongly disagree, 5=strongly agree)	Attitudes about the performance of the Forest Service (1=very unfavorable, 5=very favorable)
20. Inform public about recreation concerns (safety, trail etiquette, etc.)	4.53 *0.93*[a]	4.50 *0.93*	3.89 *1.31*
21. Inform public on potential environmental impacts of uses	4.39 *1.02*	4.48 *1.00*	3.50 *1.33*
22. Inform public on economic value from developing natural resources	4.02 *1.30*	4.08 *1.17*	3.40 *1.38*
23. Encourage collaboration between groups to share information	4.15 *1.20*	4.15 *1.17*	3.72 *1.22*

[a] Standard deviation

Strategic Plan Objective 3.d: Broaden the participation of less traditional research groups in research and technical assistance programs.

The final objective under the goal of providing scientific and technical assistance is to broaden participation in research and technical assistance programs. Table 14 shows the VOBA results for this Strategic Plan objective.

The public feels that it is an important objective for public land managers to encourage collaboration between groups, with no difference between non-metropolitan and metropolitan residents. Easterners (both non-metropolitan and metropolitan) are more supportive of this objective. The public also feels that this is an important role for the USDA Forest Service. Agency performance is rated as somewhat favorable.

Table 14. National mean scores for Strategic Plan Objective 3.d. Broaden the participation of less traditional research groups in research and technical assistance programs. Survey item is from the VOBA module of the National Survey on Recreation and the Environment.

Survey item number and statement	Objectives for the management of forests and grasslands (1=not at all important, 5=very important)	Beliefs about the role of the USDA Forest Service (1=strongly disagree, 5=strongly agree)	Attitudes about the performance of the Forest Service (1=very unfavorable, 5=very favorable)
23. Encourage collaboration between groups to share information	4.15 *1.20*[a]	4.15 *1.17*	3.72 *1.22*

[a] Standard deviation

Goal 4: Effective Public Service: Ensure the acquisition and use of an appropriate corporate infrastructure to enable the efficient delivery of a variety of uses.

The USDA Forest Service will ensure the acquisition and use of an appropriate corporate infrastructure to enable the efficient delivery of a variety of uses. To meet the goal of effective public service the USDA Forest Service has described six objectives. Three of these can be addressed using the VOBA results.

Strategic Plan Objective 4.a: Improve financial management to achieve fiscal accountability.

The first objective is to improve the financial management to achieve fiscal accountability (table 15).

One of the issues associated with a lack of fiscal accountability has been the subsidization of extractive uses of public lands. The public does not support the continuation of such subsidies, as indicated by the fact that they disagree somewhat with the values in table 15. It is interesting to note that those in the non-metropolitan West are *less* supportive of resource development subsidies than those in the non-metropolitan East. Similarly surprising, metropolitan westerners are more in favor of continuing such subsidies than are non-metropolitan westerners.

Table 15. National mean scores for Strategic Plan Objective 4.a.: Improve financial management to achieve fiscal accountability. Survey item is from the VOBA module of the National Survey on Recreation and the Environment.

Survey item number and statement	Full sample	Values with respect to forests and grasslands (1=strongly disagree, 5=strongly agree)							
		East		West		Non-metro		Metro	
		Non-metro	Metro	Non-metro	Metro	East	West	East	West
24. The Federal government should subsidize development and leasing of public lands	2.32 *1.58*[a]	2.33 *1.15*	2.28 *1.73*	2.09 *0.98*	2.42[**] *1.95*	2.33 *1.15*	2.09[*] *0.98*	2.28 *1.73*	2.42 *1.95*

[a] Standard deviation
[*], [**], [***] Mean difference is significant at $\alpha = 0.05, 0.01, 0.001$

Strategic Plan Objective 4.b: Improve the safety and economy of USDA Forest Service roads, trails, facilities, and operations and provide greater security for the public and employees.

A second objective supporting the improvement of public service is to increase the safety and economy of USDA Forest Service facilities (see table 16).

Table 16. National mean scores for Strategic Plan Objective 4.b.: Improve the safety and economy of USDA Forest Service roads, trails, facilities, and operations and provide greater security for the public and employees. Survey items are from the VOBA module of the National Survey on Recreation and the Environment.

Survey item number and statement	Objectives for the management of forests and grasslands (1=not at all important, 5=very important)	Beliefs about the role of the USDA Forest Service (1=strongly disagree, 5=strongly agree)	Attitudes about the performance of the Forest Service (1=very unfavorable, 5=very favorable)
19. Develop volunteer programs to improve facilities (trails, etc.)	4.18 *1.13*[a]	4.22 *1.09*	3.79 *1.21*
20. Inform public about recreation concerns (safety, trail etiquette, etc.)	4.53 *0.93*	4.50 *0.93*	3.89 *1.31*
29. Increasing law enforcement on National Forests and Grasslands	4.01 *1.21*	4.01 *1.26*	3.85 *1.27*

[a] Standard deviation

The public finds the development of volunteer programs to assist with facilities maintenance to be an important objective for public land management. The strongest support for this objective appears in the West. The public also believes that the development of such programs is an important role for the USDA Forest Service. Agency performance in this area is seen as somewhat favorable.

The need to inform the public about recreation concerns such as safety is seen as an important objective, especially by metropolitan residents. This is also seen as an important role for the agency. The USDA Forest Service is seen as doing a somewhat favorable job in providing information about recreation concerns.

Furthermore, the public sees an increase in law enforcement on National Forest lands as an important objective and as an important role for the agency. In the west, metropolitan residents are more supportive of increased law enforcement. Again, the agency performance is rated as somewhat favorable by all groups.

Strategic Plan Objective 4.f: Provide appropriate access to National Forest System lands and ensure nondiscrimination in the delivery of all USDA Forest Service programs.

Providing access is the final objective under the goal of improving public service. VOBA results for this objective are presented in table 17.

The public finds the preservation of local cultural uses to be a somewhat important objective for public lands. This is also seen as an appropriate role for the USDA Forest Service, with the strongest support in the West coming from metropolitan areas. Agency performance is viewed as somewhat favorable.

Providing natural resources to dependent communities is seen as a somewhat important objective, while relaxing the permitting process is seen as slightly unimportant. Those in the non-metropolitan West show the strongest support for providing resources to dependent communities, while among metropolitan dwellers, easterners are more supportive than

Table 17. National mean scores for Strategic Plan Objective 4.f.: Provide appropriate access to National Forest System lands and ensure nondiscrimination in the delivery of all USDA Forest Service programs. Survey items are from the VOBA module of the National Survey on Recreation and the Environment.

Survey item number and statement	Objectives for the management of forests and grasslands (1=not at all important, 5=very important)			Beliefs about the role of the USDA Forest Service (1=strongly disagree, 5=strongly agree)			Attitudes about the performance of the Forest Service (1=very unfavorable, 5=very favorable)		
	Full sample	Non-metro	Metro	Full sample	Non-metro	Metro	Full sample	Non-metro	Metro
11. Preserve cultural uses	3.82	3.76	3.84	3.75	3.65	3.78	3.39	3.53	3.36[*]
	1.38[a]	1.00	1.57	1.36	1.06	1.52	1.37	0.96	1.58
12. Provide natural resources to dependent communities	3.60	3.69	3.58	3.29	3.31	3.29	3.45	3.32	3.48[*]
	1.39	0.99	1.59	1.39	1.00	1.58	1.30	0.96	1.47
15. Make the permitting process easier for established uses	2.89	2.95	2.88	2.90	2.94	2.89	3.05	3.19	3.01[*]
	1.55	1.12	1.77	1.59	1.14	1.83	1.43	1.05	1.65
17. Expand commercial recreation	3.04	3.08	3.04	3.25	3.10	3.28	3.45	3.32	3.48[*]
	1.45	1.01	1.67	1.53	1.04	1.76	1.24	0.95	1.40
25. Allow for diverse uses	4.07	4.04	4.08	4.02	4.07	4.01	3.73	3.73	3.73
	1.18	0.91	1.30	1.14	0.86	1.29	1.18	0.86	1.36

[a] Standard deviation
*, **, *** Mean difference is significant at α = 0.05, 0.01, 0.001

westerners. Respondents believe that providing resources to dependent communities is a somewhat appropriate role for the agency, while relaxing the permitting process is a slightly unimportant role. Again, in the West, those in non-metropolitan areas are most supportive of the role of providing resources. Agency performance is at least slightly favorable for these objectives.

The expansion of commercial recreation is a slightly important objective. This is seen as a slightly important role for the USDA Forest Service. The agency's performance is rated as somewhat favorable. Overall, allowing for diverse uses of public lands is seen as an important objective by most groups. This is an important role for the agency, who is viewed as doing a somewhat favorable job.

C. Survey Results for the Public Lands Values (PLV)

This section presents the items in the values section grouped in a manner that is consistent with factor analysis conducted on prior data sets. Specifically, these items have been grouped into two subsets: Socially Responsible Individual Values and Socially Responsible Management Values. A subset of NSRE respondents were asked the full set of values items. This subset was then used to conduct factor analysis, which confirmed the existence of the two factors. The full demographic breakdowns for the Public Lands Values can be found in Appendix H.

Respondents were asked to rate their values for public lands (see Appendix B for the 25-item scale). This scale has been used in numerous contexts related to public land studies both at the regional and at the national level. The objective of this scale is to identify and better understand the basic underlying values that guide public land and natural resource use, management, and conservation. Based on numerous pretests, it has been determined that the values scale items can be divided into two subsets, with 17 items grouped under the heading of Socially Responsible Individual Values (SRIV) and eight items grouped under the heading of Socially Responsible Management Values (SRMV). As the name indicates, the set of values that loaded on SRIV provide information about a respondent's personal values related to public lands. A lower group mean indicates a more instrumental (or anthropocentric) view of nature, while a higher group mean implies a more biocentric perspective. The values grouped under the SRMV heading provide information about a respondent's views on how public lands should be managed. Here a lower group mean indicates a strong conservation/preservation ethic, while a higher group mean implies stronger support for resource development/consumption.

The means for each of the two values factors are 4.16 out of 5.0 for SRIV and 2.94 out of 5.0 for SRMV. Individual values scale item means are reported in table 18.

These SRIV results indicate that the public has a strong orientation toward environmental protection. The SRMV results indicate that the public holds a moderately strong conservation/preservation ethic, but that they also support some degree of development. For example, the NSRE score of 4.00 on SRMV item 4 indicates that the general public is interested in the option value of forests and grasslands, e.g., they want to preserve the option to utilize forests and grasslands in the future.

Table 18. National mean scores for individual values scale items. Items are from the VOBA module of the National Survey on Recreation and the Environment.

A. Group 1 - Socially Responsible Individual Values	Mean
1. People should be more concerned about how public lands are used	4.75[a] *0.72[b]*
2. Natural resources must be preserved, even if some people must do without	4.14 *1.22*
3. Consumers should be interested in environmental consequences of purchases	4.47 *1.05*
4. I would be willing to sign a petition for an environmental cause	4.03 *1.43*
5r.[c] The whole pollution issue *has* upset me, I feel it's *not* overrated (r)	3.68 *1.51*
6. If we could just get by with less, more for future generations	3.99 *1.35*
7. Manufacturers should be encouraged to use recycled materials	4.69 *0.82*
8. Future generations should be as important as current in public lands decisions	4.52 *0.97*
9. I would pay $5 more to use public lands for recreation	3.49 *1.60*
10. People should urge friends to limit use of scarce resources	4.14 *1.25*
11. I am glad there are National Forests even if I never see them	4.66 *0.91*
12. People can think public lands are valuable even if they don't go there	4.46 *1.07*
13. I am willing to stop buying from polluting companies	3.95 *1.35*
14. I am willing to make personal sacrifices to slow pollution₎	4.44 *1.05*
15. Forests have a right to exist for their own sake	4.11 *1.28*
16. Wildlife, plants, and humans have equal rights	4.28 *1.26*
17. Donating time or money to worthy causes is important to me	4.25 *1.05*
Group Mean	**4.24**

B. Group 2 - Socially Responsible Management Values	Mean
1. We should actively harvest more trees for larger human population	2.88[a] *1.77[b]*
2. The most important role for public lands is providing jobs, income for locals	3.23 *1.53*
3. The decision to develop resources should be made mostly on economic grounds	2.92 *1.51*
4. The main reason for maintaining resources now is to develop in future	4.00 *1.39*

continued on next page

Table 18. *Continued.*

5. I think public land managers are doing an adequate job of protecting natural resources	3.18 *1.31*
6. The primary use of forests should be for products useful for humans	2.95 *1.64*
7. The Federal government should subsidize development and leasing of public lands	2.32 *1.58*
8. The government has better places to spend money than on strong conservation program	2.33 *1.48*
Group Mean	**2.98**

[a] 1=strongly disagree, 5=strongly agree
[b] Standard deviation
[c] Values statement 5 has been reverse scored in order to calculate a group mean. See Appendix A for an explanation of reverse scoring

D. Survey Objectives, Beliefs, and Attitudes Grouped by the Focus Group's Strategic Level Objectives

This subsection presents the survey results for the objectives, beliefs and attitudes data grouped according to overarching strategic objectives expressed by the public in the focus groups. These overarching objectives are: Access, Preservation/Conservation, Economic Development, Education, and Natural Resource Management. Results organized according to these strategic level objectives and broken down by demographic subsets are presented in Appendix I.

Based upon the results of the focus group interviews, an objectives hierarchy was constructed for each group. These hierarchies indicated what each group or individual was attempting to achieve and how they would achieve each goal or objective. These objectives ranged from the very abstract strategic level to the more focused or concrete means level. The means level objectives are at the bottom of the hierarchy while the strategic objective is at the top. Fundamental objectives between the means level and the strategic level completed the hierarchies. (See Figure 1 on page 5.)

Each of the objectives hierarchies was confirmed with its respective group so as to ensure that it accurately reflected their goals and objectives. A combined objectives hierarchy was then constructed that included all the objectives stated by each group or individual interviewed. The result was a hierarchy that covered five strategic level objectives related to access, preservation/conservation, commodity development, education and natural resource management. These five strategic level objectives were supported by 30 higher-level fundamental objectives.

The 30 fundamental level objectives were used to develop the 30 objectives statements used in the NSRE survey. The 30 objectives statements were divided into five groups based upon the strategic level objectives that the focus groups had identified. During the telephone interviews each respondent was asked one statement from each of the five strategic level groups in order to obtain a statistically valid sample for each statement and for each strategic level group. The specific objectives hierarchy for the NSRE is presented in table 19.

Table 19. Objectives hierarchy for the NSRE.

Strategic Objective 1: Access

Fundamental Objective 1	Expanding access for motorized off-highway vehicles on forests and grasslands (for example snowmobiling or 4-wheel driving).
Fundamental Objective 2	Developing and maintaining continuous trail systems that cross both public and private land for motorized vehicles such as snowmobiles or ATVs.
Fundamental Objective 3	Developing and maintaining continuous trail systems that cross both public and private land for non-motorized recreation such as hiking and cross-country skiing.
Fundamental Objective 4	Designating some existing recreation trails for specific uses (for example, creating separate trails for snowmobiling and cross-country skiing, or for mountain biking and horseback riding).
Fundamental Objective 5	Developing new paved roads on forests and grasslands for access for cars and recreational vehicles
Fundamental Objective 6	Designating more wilderness areas on public land that stops access for development and motorized uses.

Strategic Objective 2: Preservation/Conservation

Fundamental Objective 7	Conserving and protecting forests and grasslands that are the source of our water resources, such as streams, lakes, and watershed areas.
Fundamental Objective 8	Preserving the natural resources of forests and grasslands through such policies as no timber harvesting or mining.
Fundamental Objective 9	Protecting ecosystems and wildlife habitats.
Fundamental Objective 10	Preserving the ability to have a wilderness experience on forests and grasslands.
Fundamental Objective 11	Preserving the cultural uses of forests and grasslands by Native Americans and Native Hispanics such as firewood gathering, herb/berry/plant gathering, and ceremonial access.

Strategic Objective 3: Economic Development

Fundamental Objective 12	Providing natural resources from forests and grasslands to support communities depending on grazing, mining, or timber harvesting.
Fundamental Objective 13	Restricting mineral development on forests and grasslands.
Fundamental Objective 14	Restricting timber harvesting and grazing on forests and grasslands.
Fundamental Objective 15	Making the permitting process easier for some established uses of forests and grasslands such as grazing, mining, and commercial recreation.
Fundamental Objective 16	Developing a national policy that guides natural resource development of all kinds (for example, specifies levels of extraction and regulates environmental impacts).
Fundamental Objective 17	Expanding commercial recreation on forests and grasslands (for example, ski areas, guide services, or outfitters).

continued on next page

Table 19. *Continued.*

Strategic Objective 4: Education	
Fundamental Objective 18	Developing volunteer programs to improve forests and grasslands (for example, planting trees, or improving water quality).
Fundamental Objective 19	Developing volunteer programs to maintain trails and facilities on forests and grasslands (for example, trail maintenance or campground maintenance).
Fundamental Objective 20	Informing the public about recreation concerns on forests and grasslands such as safety, trail etiquette, and respect for wildlife.
Fundamental Objective 21	Informing the public on the potential environmental impacts of all uses associated with forests and grasslands.
Fundamental Objective 22	Informing the public on the economic value received by developing our natural resources.
Fundamental Objective 23	Encouraging collaboration between groups in order to share information concerning uses of forests and grasslands.
Strategic Objective 5: Natural Resource Management	
Fundamental Objective 24	Using public advisory committees to advise on public land management issues.
Fundamental Objective 25	Allowing for diverse uses of forests and grasslands such as grazing, recreation, and wildlife habitat.
Fundamental Objective 26	Making management decisions concerning the use of forests and grasslands at the local level rather than at the national level.
Fundamental Objective 27	Increasing the total number of acres in the public land system.
Fundamental Objective 28	Paying an entry fee that goes to support public land.
Fundamental Objective 29	Increasing law enforcement efforts by public land agencies on public lands.
Fundamental Objective 30	Allowing public land managers to trade public lands for private lands (for example, to eliminate private property within public land boundaries, or to acquire unique areas of land).

Access

Items dealing with access are those that refer to expanding off-highway motorized access and paved roads, increasing trails for motorized and non-motorized recreation, and designating separate trails for specific uses. The results for this group are in table 20.

Noteworthy is the difference the between how respondents feel about motorized access and how they feel about non-motorized access. Respondents view the objectives of expansion of off-highway motorized access and the development of new paved roads to be somewhat unimportant objectives and the provision of trails for motorized access to be slightly unimportant. Contrast this with the provision of non-motorized access, which is viewed as a somewhat important objective for respondents. The public's beliefs about these activities as roles for the USDA Forest Service parallel their objectives, with motorized access being a slightly or somewhat unimportant role and non-motorized access being a somewhat important role.

Motorized and non-motorized activities can often come into conflict, and the public sees the designation of trails for specific uses as a somewhat important objective and a somewhat important role for the agency. Wilderness designation also impacts access and the public feels that additional designations are a somewhat important objective.

Table 20. National mean scores for access. Survey items are from the VOBA module of the National Survey on Recreation and the Environment.

Survey item number and statement	**Objectives** for the management of forests and grasslands *(1=not at all important, 5=very important)*	**Beliefs** about the role of the USDA Forest Service *(1=strongly disagree, 5=strongly agree)*	**Attitudes** about the performance of the Forest Service *(1=very unfavorable, 5=very favorable)*
1. Expand off-highway motorized access	2.41 *1.49[a]*	2.52 *1.50*	2.97 *1.45*
2. Increase trails for motorized vehicles	2.82 *1.47*	2.98 *1.51*	3.25 *1.38*
3. Increase trails for non-motorized recreation	3.75 *1.33*	3.71 *1.37*	3.59 *1.27*
4. Designate trails for specific uses	3.59 *1.43*	3.94 *1.25*	3.61 *1.33*
5. Develop more paved roads	2.62 *1.49*	2.70 *1.57*	3.19 *1.43*
6. Designate more wilderness to stop development & motorized access	3.84 *1.41*	3.66 *1.46*	3.45 *1.34*

[a] Standard deviation

Some regional differences do occur. In the East, non-metropolitan residents are more likely to support the expansion of trails for motorized vehicles. In the West, it is metropolitan residents who are most in favor of trails for non-motorized recreation and the development of new paved roads. In the West, non-metropolitan residents see the designation of specific-use trails to be a more important objective than do their metropolitan counterparts. Among metropolitan residents, westerners are slightly more supportive of the expansion of off-highway motorized access.

Agency performance is rated as at least slightly favorable, with the exception of the expansion of off-highway motorized access, which is rated as slightly unfavorable. This may indicate that the public sees the agency as providing too much of this type of access, since they find this to be both an unimportant role and an unimportant objective.

Preservation/Conservation

These next items represent the strategic level objective dealing with preservation of wildlands and conservation of natural resources. Table 21 shows the results for this group.

It is interesting to note that the public feels that the conservation and protection of watersheds is an important objective, consistent with the USDA Forest Service Organic Act. Also, important objectives for the public are the preservation of natural resources through policies that restrict commodity uses, protection of ecosystems and wildlife habitat, and preservation of the ability to enjoy a "wilderness" experience. A somewhat important objective is the preservation of local cultural uses.

As with the "Access" items, the public's beliefs about the role the USDA Forest Service should play in fulfilling these objectives mirror their opinions about the objectives importance. And as before, the agency performance is rated as somewhat favorable for most items.

Metropolitan residents in both the East and West are more in favor of policies that restrict timber and mineral extraction. Metropolitan dwellers in the West see protection of ecosystems as a more important objective than do non-metropolitan westerners. Looking at non-metropolitan residents, those in the East are more likely to favor the protection of ecosystems. Metropolitan dwellers in the East place a higher rating on the preservation of a "wilderness experience" than western metropolitan residents.

Table 21. National mean scores for preservation/conservation. Survey items are from the VOBA module of the National Survey on Recreation and the Environment.

Survey item number and statement	Objectives for the management of forests and grasslands (1=not at all important, 5=very important)	Beliefs about the role of the USDA Forest Service (1=strongly disagree, 5=strongly agree)	Attitudes about the performance of the Forest Service (1=very unfavorable, 5=very favorable)
7. Conserve and protect watersheds	4.73 *0.76[a]*	4.61 *0.83*	3.91 *1.17*
8. Preserve natural resources through policies such as no timber, no mining	4.22 *1.23*	4.21 *1.27*	3.65 *1.31*
9. Protecting ecosystems and wildlife habitat	4.58 *0.92*	4.53 *0.98*	3.90 *1.16*
10. Preserve "wilderness" experience	4.15 *1.28*	4.22 *1.14*	3.88 *1.10*
11. Preserve local cultural uses	3.82 *1.38*	3.75 *1.36*	3.39 *1.37*

[a] Standard deviation

Economic Development

Development of economically valuable resources is the third strategic level objective. VOBA items that are related to this objective are presented in table 22.

These items deal with commodity development and commercial uses of public lands. All the items in this objective are seen as at least slightly important objectives, with the exception of making the permitting process easier, which is viewed as slightly unimportant. The development of a national policy for natural resource development is seen as the most important of the items in this objective. The public sees the restriction of mineral development and of timber harvest and grazing as being more important than the provision of natural resources to dependent communities (although this is still seen as somewhat important). Least important are the easing of permit processes and the expansion of commercial recreation. Residents of metropolitan areas are more in favor of a national policy for natural resource development than are non-metropolitan residents in both the East and West.

Table 22. National mean scores for economic development. Survey items are from the VOBA module of the National Survey on Recreation and the Environment.

Survey item number and statement	Objectives for the management of forests and grasslands (1=not at all important, 5=very important)	Beliefs about the role of the USDA Forest Service (1=strongly disagree, 5=strongly agree)	Attitudes about the performance of the Forest Service (1=very unfavorable, 5=very favorable)
12. Provide natural resources to dependent communities	3.60 *1.39[a]*	3.29 *1.39*	3.45 *1.30*
13. Restrict development of minerals	3.96 *1.42*	3.95 *1.43*	3.30 *1.50*
14. Restrict timber harvest and grazing	3.99 *1.27*	3.94 *1.34*	3.50 *1.36*
15. Make the permitting process easier for established uses	2.89 *1.55*	2.90 *1.59*	3.05 *1.43*
16. Develop a national policy for natural resource development	4.26 *1.23*	4.21 *1.19*	3.51 *1.26*
17. Expand commercial recreation	3.04 *1.45*	3.25 *1.53*	3.45 *1.24*

[a] Standard deviation

Education

Respondents express a fourth strategic level objective concerning education and the distribution of information. These results are presented in table 23.

All of the items in this strategic level objective are considered to be important objectives by the public. The public also feels that fulfilling these objectives are important roles for the USDA Forest Service to play. Agency performance in this area is rated as somewhat favorable by the public.

There are few striking demographic differences with this strategic level objective. One that stands out is the fact that metropolitan residents (both easterners and westerners) are more in favor of informing the public about potential environmental impacts of uses of public lands than are non-metropolitan residents. Among residents of metropolitan areas, those in the East see this objective as more important than metropolitan westerner.

Table 23. National mean scores for education. Survey items are from the VOBA module of the National Survey on Recreation and the Environment.

Survey item number and statement	Objectives for the management of forests and grasslands (1=not at all important, 5=very important)	Beliefs about the role of the USDA Forest Service (1=strongly disagree, 5=strongly agree)	Attitudes about the performance of the Forest Service (1=very unfavorable, 5=very favorable)
18. Develop volunteer programs to improve land (tree planting, etc.)	4.60 0.86[a]	4.46 1.03	3.85 1.25
19. Develop volunteer programs to improve facilities (trails etc.)	4.18 1.13	4.22 1.09	3.79 1.21
20. Inform the public about recreation concerns (safety, trail etiquette, etc.)	4.53 0.93	4.50 0.93	3.89 1.31
21. Inform the public on potential environmental impacts of uses	4.39 1.02	4.48 1.00	3.50 1.33
22. Inform the public on economic value from developing natural resources	4.02 1.30	4.08 1.17	3.40 1.38
23. Encourage collaboration between groups to share information	4.15 1.20	4.15 1.17	3.72 1.22

[a] Standard deviation

Natural Resource Management

The management of natural resources is also an important strategic level objective expressed by focus group participants. Items that relate to this objective appear in table 24.

Public input into natural resource management decisions is of interest to survey respondents, with both the use of public advisory committees and local level decision making being somewhat important objectives. Allowing for diverse uses is also an important objective. These items are also seen as appropriate roles for the USDA Forest Service.

Increasing the total number of acres in the National Forest and Grassland system is somewhat important to the public, while land trades to eliminate inholdings is slightly important. Both of these items are seen as somewhat important roles for the agency.

Increasing law enforcement on USDA Forest Service lands is an important objective. Collection of an entry fee is somewhat important as well. Agency performance in each of these areas is at least slightly favorable.

Table 24. National mean scores for natural resource management. Survey items are from the VOBA module of the National Survey on Recreation and the Environment.

Survey item number and statement	**Objectives** for the management of forests and grasslands *(1=not at all important, 5=very important)*	**Beliefs** about the role of the USDA Forest Service *(1=strongly disagree, 5=strongly agree)*	**Attitudes** about the performance of the Forest Service *(1=very unfavorable, 5=very favorable)*
24. Using public advisory committees to advise on management issues	3.90 *1.20[a]*	3.84 *1.25*	3.36 *1.26*
25. Allow for diverse uses	4.07 *1.18*	4.02 *1.14*	3.73 *1.18*
26. Make management decisions at the local level (rather than national)	3.93 *1.22*	3.88 *1.34*	3.49 *1.32*
27. Increase total number of acres in the National Forest and Grassland system	3.81 *1.42*	3.95 *1.47*	3.52 *1.45*
28. Collect an entry fee to support National Forests and Grasslands	3.66 *1.36*	3.69 *1.43*	3.61 *1.36*
29. Increasing law enforcement on National Forests and Grasslands	4.01 *1.21*	4.01 *1.26*	3.85 *1.27*
30. Trade public lands for private to eliminate inholdings, acquire unique lands	3.05 *1.48*	3.22 *1.49*	3.22 *1.27*

[a] Standard deviation

E. Survey Values, Objectives, Beliefs, and Attitudes Linked to the Level of Respondent Familiarity With the USDA Forest Service

This section contains results for 5,064 respondents who were asked questions designed to ascertain their familiarity with the USDA Forest Service. These questions were a set of five statements pertaining to management roles and characteristics of the USDA Forest Service. Respondents were asked to give "yes/no" answers (essentially "true/false"). The exact wording of the familiarity questions can be found in Appendix F.

Responses to these five questions were coded to indicate whether the respondent gave a correct answer. The total number of correct answers was then tallied, and finally a variable indicating the level of familiarity was constructed based on the number of correct answers. A respondent was classified as having "high familiarity" if he/she correctly answered four or

Table 25. Demographics by familiarity with the USDA Forest Service. Results are from the VOBA module of the National Survey on Recreation and the Environment.

	Low familiarity	Moderate familiarity	High familiarity	Significant difference Source of difference
Age	40.58 24.28[a]	44.08 17.77	44.37 13.63	*** those with low familiarity are different from rest
Income	$44,966 $40,750	$61,027 $22,9363	$62,445 $50,806	
Education level [b]	2.14 1.43	2.68 1.22	3.17 1.05	*** all

[a] Standard deviation

[b] Education categories: 1=less than high school, 2=high school graduate, 3=some college, 4=bachelor s degree, 5=post graduate degree

*, **, *** Mean difference is significant at α = 0.05, 0.01, 0.001

five of the questions. Respondents who correctly answered two or three were classified as having "moderate familiarity" and those who answered one or fewer were classified as having "low familiarity." Demographic descriptions of the three familiarity groups are in table 25.

These results will be presented using the same item groupings as above: first by Strategic Plan objectives, then the Public Lands Values, and finally by Strategic Level Objectives.

1. VOBA Results Linked to USDA Forest Service Strategic Plan Objectives by Familiarity

Generally, when a statistically significant difference between respondents with high familiarity and those with low or moderate familiarity is found, it is the case that those with lower familiarity have the higher response. This is especially noteworthy when looking at the items dealing with attitudes about the performance of the USDA Forest Service. It would appear that those less familiar with the agency are more likely to give the agency higher performance ratings and to rate all objectives as important for themselves and as agency roles, "yeah-saying" if you will.

Goal 1: Ecosystem Health: Promote ecosystem health and conservation using a collaborative approach to sustain the Nation's forests, grasslands, and watersheds.

The USDA Forest Service will promote ecosystem health and conservation using a collaborative approach to sustain the Nation's forests, grasslands, and watersheds. Supporting this goal are three specific objectives.

Strategic Plan Objective 1.a: Improve and protect watershed conditions to provide the water quality and quantity and the soil productivity necessary to support ecological functions and intended beneficial water uses.

The first objective supporting ecosystem health deals with protecting water and soil resources. The public responses for the items in the VOBA that support this objective, categorized by respondent familiarity, are in table 26.

The development of volunteer programs to improve the land is most strongly supported by those with a moderate level of agency familiarity, and the least support is found among those with a lower level of familiarity (however, these respondents give the most favorable rating of agency performance).

Table 26. Mean scores by respondent familiarity for Strategic Plan Objective 1.a.: Improve and protect watershed conditions to provide the water quality and quantity and the soil productivity necessary to support ecological functions and intended beneficial water uses. Survey items are from the VOBA module of the National Survey on Recreation and the Environment.

Survey item number and statement	Objectives for the management of forests and grasslands (1=not at all important, 5=very important)				Signif. diff. Source of diff.
	Full sample N=5,064	Low familiarity N=942	Moderate familiarity N=3,533	High familiarity N=589	
7. Conserve and protect watersheds	4.73 0.76[a]	4.70 0.87	4.72 0.71	4.74 0.68	
18. Develop volunteer programs to improve the land (tree planting, etc.)	4.60 0.86	4.41 1.13	4.62 0.80	4.50 0.85	* L-M[b]

continued on next page

Table 26. *Continued.*

| Survey item number and statement | Beliefs about the role of the USDA Forest Service (1=strongly disagree, 5=strongly agree) | | | | |
	Full sample N=5,064	Low familiarity N=942	Moderate familiarity N=3,533	High familiarity N=589	Signif. diff. Source of diff.
7. Conserve and protect watersheds	4.61 *0.83*	4.60 *0.88*	4.62 *0.79*	4.61 *0.64*	
18. Develop volunteer programs to improve the land (tree planting, etc.)	4.46 *1.03*	4.41 *1.30*	4.54 *0.85*	4.52 *0.83*	

| Survey item number and statement | Attitudes about the performance of the Forest Service (1=very unfavorable, 5=very favorable) | | | | |
	Full sample N=5,064	Low familiarity N=942	Moderate familiarity N=3,533	High familiarity N=589	Signif. diff. Source of diff.
7. Conserve and protect watersheds	3.91 *1.17*	3.97 *1.37*	3.95 *1.06*	3.67 *1.03*	
18. Develop volunteer programs to improve the land (tree planting, etc.)	3.85 *1.25*	4.16 *1.41*	3.82 *1.17*	3.30 *1.13*	*** L-M, H; M-H

[a] Standard deviation
[b] For example, L-M indicates that the source of the statistically significant difference lies in the differences between the responses of those with low agency familiarity and those with moderate familiarity
*, **, *** Mean difference is significant at = 0.05, 0.01, 0.001

Strategic Plan Objective 1.b: Provide ecological conditions to sustain viable populations of native and desired nonnative species and to achieve objectives for Management Indicator Species (MIS)/focal species.

Habitat protection is the second objective under the goal of ecosystem health. Responses by familiarity are presented in table 27.

Here it is interesting to note that those respondents with lower agency familiarity are the least in favor of restricting mineral development. When asked about policies that preserve natural resources through no timber harvesting and no mining, however, those with the highest degree of agency familiarity are least in favor.

Table 27. Mean scores by respondent familiarity for Strategic Plan Objective 1.b.: Provide ecological conditions to sustain viable populations of native and desired nonnative species and to achieve objectives for Management Indicator Species (MIS)/focal species. Survey items are from the VOBA module of the National Survey on Recreation and the Environment.

| Survey item number and statement | Objectives for the management of forests and grasslands (1= not at all important, 5= very important) | | | | |
	Full sample N=5,064	Low familiarity N=942	Moderate familiarity N=3,533	High familiarity N=589	Signif. diff. Source of diff.
6. Designate more wilderness to stop development & motorized access	3.84 *1.41[a]*	3.65 *1.67*	3.83 *1.30*	3.99 *1.14*	
8. Preserve natural resources through policies such as no timber, no mining	4.22 *1.23*	4.18 *1.46*	4.31 *1.07*	3.99 *1.14*	* M-H

continued on next page

USDA Forest Service RMRS GTR-95. 2002

Table 27. *Continued.*

9. Protect ecosystems and wildlife habitat	4.58 *0.92*	4.48 *1.19*	4.59 *0.84*	4.55 *0.88*	
13. Restrict development of minerals	3.96 *1.42*	3.51 *1.87*	4.14 *1.19*	3.97 *1.08*	*** L-M, H
14. Restrict timber harvest and grazing	3.99 *1.27*	4.00 *1.41*	4.05 *1.18*	3.76 *1.22*	

Beliefs
about the role of the USDA Forest Service
(1= strongly disagree, 5= strongly agree)

Survey item number and statement	Full sample N=5,064	Low familiarity N=942	Moderate familiarity N=3,533	High familiarity N=589	Signif. diff. Source of diff.
6. Designate more wilderness to stop development & motorized access	3.66 *1.46*	3.57 *1.50*	3.71 *1.39*	3.61 *1.39*	
8. Preserve natural resources through policies such as no timber, no mining	4.21 *1.27*	4.23 *1.31*	4.22 *1.09*	3.98 *1.22*	
9. Protect ecosystems and wildlife habitat	4.53 *0.98*	4.40 *1.26*	4.55 *0.91*	4.23 *1.24*	* M-H
13. Restrict development of minerals	3.95 *1.43*	3.57 *1.81*	4.01 *1.34*	3.96 *1.22*	** L-M
14. Restrict timber harvest and grazing	3.94 *1.34*	3.90 *1.41*	4.08 *1.21*	3.77 *1.26*	

Attitudes
about the performance of the Forest Service
(1= very unfavorable, 5= very favorable)

Survey item number and statement	Full sample N=5,064	Low familiarity N=942	Moderate familiarity N=3,533	High familiarity N=589	Signif. diff. Source of diff.
6. Designate more wilderness to stop development & motorized access	3.45 *1.34*	3.48 *1.35*	3.30 *1.30*	3.16 *1.21*	
8. Preserve natural resources through policies such as no timber, no mining	3.65 *1.31*	3.81 *1.33*	3.74 *1.21*	3.56 *1.13*	
9. Protect ecosystems and wildlife habitat	3.90 *1.16*	4.13 *1.30*	4.02 *1.01*	3.52 *1.14*	*** H-L, M[b]
13. Restrict development of minerals	3.30 *1.50*	3.25 *1.58*	3.33 *1.49*	3.08 *1.28*	
14. Restrict timber harvest and grazing	3.50 *1.36*	3.63 *1.50*	3.42 *1.28*	3.09 *1.15*	* L-H

[a] Standard deviation
[b] For example, L-M indicates that the source of the statistically significant difference lies in the differences between the responses of those with low agency familiarity and those with moderate familiarity
*, **, *** Mean difference is significant at = 0.05, 0.01, 0.001

Strategic Plan Objective 1.c: Increase the amount of forests and grasslands restored to or maintained in a healthy condition with reduced risk and damage from fires, insects and diseases, and invasive species.

The restoration and maintenance of ecosystem health is the third objective under ecosystem management. The results by familiarity level are shown in table 28.

The least amount of support for volunteer programs to improve the land comes from those with the lowest level of agency familiarity. And as noted above, those with the lowest familiarity with the USDA Forest Service offer the highest approval rating for these items.

Table 28. Means by respondent familiarity - Strategic Plan Objective 1.c.: Increase the amount of forests and grasslands restored to or maintained in a healthy condition with reduced risk and damage from fires, insects and diseases, and invasive species. Survey items are from the VOBA module of the National Survey on Recreation and the Environment.

Survey item number and statement	**Objectives** for the management of forests and grasslands (1= not at all important, 5= very important)				
	Full sample N=5,064	Low familiarity N=942	Moderate familiarity N=3,533	High familiarity N=589	Signif. diff. Source of diff.
9. Protect ecosystems and wildlife habitat	4.58 0.92[a]	4.48 1.19	4.59 0.84	4.55 0.88	
18. Develop volunteer programs to improve the land (tree planting, etc.)	4.60 0.86	4.41 1.13	4.62 0.80	4.50 0.85	* L-M[b]

Survey item number and statement	**Beliefs** about the role of the USDA Forest Service (1= strongly disagree, 5= strongly agree)				
	Full sample N=5,064	Low familiarity N=942	Moderate familiarity N=3,533	High familiarity N=589	Signif. diff. Source of diff.
9. Protect ecosystems and wildlife habitat	4.53 0.98	4.40 1.26	4.55 0.91	4.23 1.24	* M-H
18. Develop volunteer programs to improve the land (tree planting, etc.)	4.46 1.03	4.41 1.30	4.54 0.85	4.52 0.83	

Survey item number and statement	**Attitudes** about the performance of the Forest Service (1= very unfavorable, 5= very favorable)				
	Full sample N=5,064	Low familiarity N=942	Moderate familiarity N=3,533	High familiarity N=589	Signif. diff. Source of diff.
9. Protect ecosystems and wildlife habitat	3.90 1.16	4.13 1.30	4.02 1.01	3.52 1.14	*** H-L,M
18. Develop volunteer programs to improve the land (tree planting, etc.)	3.85 1.25	4.16 1.41	3.82 1.17	3.30 1.13	*** L-M, H; M-H

[a] Standard deviation
[b] For example, L-M indicates that the source of the statistically significant difference lies in the differences between the responses of those with low agency familiarity and those with moderate familiarity
*, **, *** Mean difference is significant at = 0.05, 0.01, 0.001

Goal 2: Multiple Benefits to People: Provide a variety of uses, values, products, and services for present and future generations by managing within the capability of sustainable ecosystems.

The second broad goal outlined in the Strategic Plan deals with the provisions of benefits to the American public. The USDA Forest Service seeks to provide for a variety of uses, values, products, and services for present and future generations by managing within the capability of sustainable ecosystems. This goal is supported by five specific objectives.

Strategic Plan Objective 2.a: Improve the capability of the Nation's forests and grasslands to provide diverse, high-quality outdoor recreation opportunities.

The first Strategic Plan objective deals with providing outdoor recreation opportunities to the public. The NSRE results by level of agency familiarity are presented in table 29.

USDA Forest Service RMRS GTR-95. 2002

Members of the public with a high degree of agency familiarity have the lowest level of support for expanded off-highway motorized recreation and adding new paved roads, while those with moderate familiarity are most in favor of trails for non-motorized recreation. Those with a high level of familiarity see the expansion of motorized access and commercial recreation as less appropriate roles for the USDA Forest Service than do those who are less familiar. Expanded commercial recreation is least supported by the group most familiar with the agency. Also, those with a higher level of agency awareness offer the least support for collecting an entrance fee for National Forests and Grasslands.

As we have seen with other items, where a difference in attitudes exists, those who are least familiar with the agency tend to rate the performance higher than those who are more familiar, with the exception of providing trails for non-motorized access.

Table 29. Mean scores by respondent familiarity for Strategic Plan Objective 2.a.: Improve the capability of the Nation's forests and grasslands to provide diverse, high-quality outdoor recreation opportunities. Survey items are from the VOBA module of the National Survey on Recreation and the Environment.

Survey item number and statement	**Objectives** for the management of forests and grasslands (1= not at all important, 5= very important)				
	Full sample N=5,064	Low familiarity N=942	Moderate familiarity N=3,533	High familiarity N=589	Signif. diff. Source of diff.
1. Expand off-highway motorized access	2.41 *1.49*[a]	2.41 *1.86*	2.45 *1.40*	2.04 *1.13*	
2. Trails for motorized vehicles	2.82 *1.47*	2.98 *1.51*	2.83 *1.39*	2.99 *1.49*	
3. Trails for non-motorized recreation	3.75 *1.33*	3.49 *1.36*	3.82 *1.24*	3.50 *1.13*	** L-M
4. Designate trails for specific uses	3.59 *1.43*	3.68 *1.58*	3.61 *1.37*	3.34 *1.16*	
5. Develop new paved roads	2.62 *1.49*	2.91 *1.75*	2.57 *1.39*	2.17 *1.14*	*** L-M, H
17. Expand commercial recreation	3.04 *1.45*	3.32 *1.76*	2.85 *1.25*	2.76 *1.12*	*** L-M, H
19. Develop volunteer programs to improve facilities (trails, etc.)	4.18 *1.13*	4.09 *1.47*	4.19 *1.02*	4.20 *0.88*	
20. Inform the public about recreation concerns (safety, trail etiquette, etc.)	4.53 *0.93*	4.37 *1.18*	4.57 *0.84*	4.63 *0.63*	* L-M
28. Collect an entrance fee to support National Forests and Grasslands	3.66 *1.36*	3.43 *1.76*	3.81 *1.21*	2.94 *1.23*	*** M-L, H

Survey item number and statement	**Beliefs** about the role of the USDA Forest Service (1= strongly disagree, 5= strongly agree)				
	Full sample N=5,064	Low familiarity N=942	Moderate familiarity N=3,533	High familiarity N=589	Signif. diff. Source of diff.
1. Expand off-highway motorized access	2.52 *1.50*	2.52 *1.60*	2.60 *1.43*	2.01 *1.07*	** H-L, M[b]
2. Trails for motorized vehicles	2.98 *1.51*	3.20 *1.62*	2.92 *1.45*	2.65 *1.13*	* L-H
3. Trails for non-motorized recreation	3.71 *1.37*	3.77 *1.60*	3.67 *1.34*	3.47 *1.13*	

continued on next page

Table 29. *Continued.*

Survey item number and statement	Full sample	Low familiarity	Moderate familiarity	High familiarity	Signif. diff. Source of diff.
4. Designate trails for specific uses	3.94 *1.25*	3.88 *1.50*	4.00 *1.10*	3.57 *1.10*	* M-H
5. Develop new paved roads	2.70 *1.57*	2.89 *1.83*	2.57 *1.42*	1.92 *0.97*	*** L-M, H; M-H
17. Expand commercial recreation	3.25 *1.53*	3.13 *1.59*	3.24 *1.44*	2.78 *1.26*	* M-H
19. Develop volunteer programs to improve facilities (trails, etc.)	4.22 *1.09*	4.24 *1.15*	4.28 *0.99*	4.14 *0.85*	
20. Inform the public about recreation concerns (safety, trail etiquette, etc.)	4.50 *0.93*	4.57 *0.91*	4.53 *0.91*	4.47 *0.61*	
28. Collect an entrance fee to support National Forests and Grasslands	3.69 *1.43*	3.84 *1.54*	3.74 *1.26*	3.31 *1.19*	* H-L, M

	Attitudes about the performance of the Forest Service *(1= very unfavorable, 5= very favorable)*				
Survey item number and statement	Full sample N=5,064	Low familiarity N=942	Moderate familiarity N=3,533	High familiarity N=589	Signif. diff. Source of diff.
1. Expand off-highway motorized access	2.97 *1.45*	3.16 *1.63*	2.90 *1.38*	3.10 *1.16*	
2. Trails for motorized vehicles	3.25 *1.38*	3.39 *1.48*	3.35 *1.28*	2.98 *1.01*	*
3. Trails for non-motorized recreation	3.59 *1.27*	3.47 *1.42*	3.73 *1.19*	3.24 *0.88*	* M-H
4. Designate trails for specific uses	3.61 *1.33*	3.44 *1.48*	3.71 *1.11*	3.39 *1.06*	* L-M
5. Develop new paved roads	3.19 *1.43*	3.33 *1.71*	3.18 *1.32*	3.01 *1.10*	
17. Expand commercial recreation	3.45 *1.24*	3.47 *1.37*	3.54 *1.21*	3.36 *0.99*	
19. Develop volunteer programs to improve facilities (trails, etc.)	3.79 *1.21*	4.09 *1.14*	3.82 *1.09*	3.57 *1.02*	** L-M, H
20. Inform the public about recreation concerns (safety, trail etiquette, etc.)	3.89 *1.31*	3.95 *1.48*	3.89 *1.21*	3.83 *1.05*	
28. Collect an entrance fee to support National Forests and Grasslands	3.61 *1.36*	3.85 *1.41*	3.52 *1.30*	3.24 *0.88*	* L-M, H

	Values with respect to forests and grasslands *(1= strongly disagree, 5= strongly agree)*				
Survey item number and statement	Full sample N=5,064	Low familiarity N=942	Moderate familiarity N=3,533	High familiarity N=589	Signif. diff. Source of diff.
9. I would pay $5 more to use public lands for recreation	3.49 *1.60*	3.53 *1.87*	3.51 *1.53*	3.38 *1.31*	

[a] Standard deviation
[b] For example, L-M indicates that the source of the statistically significant difference lies in the differences between the responses of those with low agency familiarity and those with moderate familiarity
*, **, *** Mean difference is significant at α = 0.05, 0.01, 0.001

Strategic Plan Objective 2.b: Improve the capability of wilderness and protected areas to sustain a desired range of benefits and values.

The second objective under "providing benefits to people" is to provide a desired range of values and benefits from wilderness areas. Table 30 shows the results by familiarity for these items.

Table 30. Mean scores by respondent familiarity for Strategic Plan Objective 2. b.: Improve the capability of wilderness and protected areas to sustain a desired range of benefits and values. Survey items are from the VOBA module of the National Survey on Recreation and the Environment.

Survey item number and statement	Objectives for the management of forests and grasslands (1= not at all important, 5= very important)				
	Full sample N=5,064	Low familiarity N=942	Moderate familiarity N=3,533	High familiarity N=589	Signif. diff. Source of diff.
6. Designate more wilderness to stop development & motorized access	3.84 1.41[a]	3.65 1.67	3.83 1.30	3.99 1.14	
8. Preserve natural resources through policies such as no timber, no mining	4.22 1.23	4.18 1.46	4.31 1.07	3.99 1.14	* M-H[b]
9. Protect ecosystems and wildlife habitat	4.58 0.92	4.48 1.19	4.59 0.84	4.55 0.88	
10. Preserve wilderness experience	4.15 1.28	3.99 1.42	4.16 1.18	3.96 1.09	

Survey item number and statement	VOBA Beliefs about the role of the USDA Forest Service (1= strongly disagree, 5= strongly agree)				
	Full sample N=5,064	Low familiarity N=942	Moderate familiarity N=3,533	High familiarity N=589	Signif. diff. Source of diff.
6. Designate more wilderness to stop development & motorized access	3.66 1.46	3.57 1.50	3.71 1.39	3.61 1.39	
8. Preserve natural resources through policies such as no timber, no mining	4.21 1.27	4.23 1.31	4.22 1.09	3.98 1.22	
9. Protect ecosystems and wildlife habitat	4.53 0.98	4.40 1.26	4.55 0.91	4.23 1.24	* M-H
10. Preserve wilderness experience	4.22 1.14	4.06 1.29	4.32 1.05	4.12 1.08	** L-M

Survey item number and statement	VOBA Attitudes about the performance of the Forest Service (1= very unfavorable, 5= very favorable)				
	Full sample N=5,064	Low familiarity N=942	Moderate familiarity N=3,533	High familiarity N=589	Signif. diff. Source of diff.
6. Designate more wilderness to stop development & motorized access	3.45 1.34	3.48 1.35	3.30 1.30	3.16 1.21	
8. Preserve natural resources through policies such as no timber, no mining	3.65 1.31	3.81 1.33	3.74 1.21	3.56 1.13	
9. Protect ecosystems and wildlife habitat	3.90 1.16	4.13 1.30	4.02 1.01	3.52 1.14	*** H-L, M
10. Preserve wilderness experience	3.88 1.10	3.80 1.25	3.93 1.06	3.72 0.85	

[a] Standard deviation
[b] For example, L-M indicates that the source of the statistically significant difference lies in the differences between the responses of those with low agency familiarity and those with moderate familiarity
*, **, *** Mean difference is significant at = 0.05, 0.01, 0.001

Respondents with lower familiarity feel that the preservation of a wilderness experience is less important as an agency activity. It is surprising that more familiar respondents feel that protection of ecosystems and wildlife habitat is less important as an agency role than do those less familiar. Agency performance is rated higher for habitat and ecosystem protection by those least familiar with the agency.

Strategic Plan Objective 2.c: Improve the capability of the Nation's forests and grasslands to provide desired sustainable levels of uses, values, products, and services.

This objective deals with the provision of products and services to the public. A breakdown of the NSRE responses by familiarity is in table 31.

Table 31. Mean scores by respondent familiarity for Strategic Plan Objective 2.c.: Improve the capability of the Nation's forests and grasslands to provide desired sustainable levels of uses, values, products, and services. Survey items are from the VOBA module of the National Survey on Recreation and the Environment.

	Objectives for the management of forests and grasslands (1= not at all important, 5= very important)				
Survey item number and statement	Full sample N=5,064	Low familiarity N=942	Moderate familiarity N=3,533	High familiarity N=589	Signif. diff. Source of diff.
11. Preserve local cultural uses	3.82 *1.38[a]*	3.85 *1.54*	3.71 *1.35*	3.73 *1.16*	
12. Provide natural resources to dependent communities	3.60 *1.39*	3.67 *1.39*	3.58 *1.39*	3.31 *1.24*	
15. Make the permitting process easier for established uses	2.89 *1.55*	3.40 *1.69*	2.81 *1.44*	2.57 *1.26*	*** L-M, H[b]
16. Develop a national policy for natural resource development	4.26 *1.23*	4.28 *1.31*	4.27 *1.12*	4.33 *1.05*	
18. Develop volunteer programs to improve the land (tree planting, etc.)	4.60 *0.86*	4.41 *1.13*	4.62 *0.80*	4.50 *0.85*	* L-M
25. Allow for diverse uses	4.07 *1.18*	3.94 *1.42*	4.15 *1.05*	3.91 *0.91*	*
27. Increase the total number of acres in the National Forest and Grassland system	3.81 *1.42*	4.02 *1.36*	3.66 *1.38*	3.59 *1.23*	* L-M

	Beliefs about the role of the USDA Forest Service (1= strongly disagree, 5= strongly agree)				
Survey item number and statement	Full sample N=5,064	Low familiarity N=942	Moderate familiarity N=3,533	High familiarity N=589	Signif. diff. Source of diff.
11. Preserve local cultural uses	3.75 *1.36*	3.89 *1.52*	3.74 *1.31*	3.50 *1.10*	* L-H
12. Provide natural resources to dependent communities	3.29 *1.39*	3.46 *1.43*	3.38 *1.31*	3.12 *1.09*	
15. Make the permitting process easier for established uses	2.90 *1.59*	2.70 *1.81*	2.77 *1.43*	2.45 *1.40*	
16. Develop a national policy for natural resource development	4.21 *1.19*	4.04 *1.48*	4.28 *1.11*	4.05 *1.00*	* L-M
18. Develop volunteer programs to improve the land (tree planting, etc.)	4.46 *1.03*	4.41 *1.30*	4.54 *0.85*	4.52 *0.83*	

continued on next page

Table 31. *Continued.*

Survey item number and statement	Full sample N=5,064	Low familiarity N=942	Moderate familiarity N=3,533	High familiarity N=589	Signif. diff. Source of diff.
25. Allow for diverse uses	4.02 *1.14*	3.96 *1.18*	4.04 *1.12*	4.13 *0.90*	
27. Increase the total number of acres in the National Forest and Grassland system	3.95 *1.47*	3.57 *1.78*	4.06 *1.26*	3.69 *1.22*	*** L-M

Survey item number and statement	Attitudes about the performance of the Forest Service (1= very unfavorable, 5= very favorable)				
	Full sample N=5,064	Low familiarity N=942	Moderate familiarity N=3,533	High familiarity N=589	Signif. diff. Source of diff.
11. Preserve local cultural uses	3.39 *1.37*	3.59 *1.25*	3.46 *1.30*	3.32 *0.98*	
12. Provide natural resources to dependent communities	3.45 *1.30*	3.21 *1.53*	3.54 *1.16*	2.90 *1.09*	*** M-L, H
15. Make the permitting process easier for established uses	3.05 *1.43*	3.08 *1.75*	3.10 *1.34*	2.60 *0.97*	
16. Develop a national policy for natural resource development	3.51 *1.26*	3.71 *1.30*	3.53 *1.17*	3.46 *1.12*	
18. Develop volunteer programs to improve the land (tree planting, etc.)	3.85 *1.25*	4.16 *1.41*	3.82 *1.17*	3.30 *1.13*	*** L-M, H; M-H
25. Allow for diverse uses	3.73 *1.18*	3.91 *1.10*	3.82 *1.10*	3.44 *0.98*	* H-L, M
27. Increase the total number of acres in the National Forest and Grassland system	3.52 *1.45*	3.61 *1.54*	3.45 *1.28*	3.00 *1.12*	* H-L, M

[a] Standard deviation
[b] For example, L-M indicates that the source of the statistically significant difference lies in the differences between the responses of those with low agency familiarity and those with moderate familiarity
[*], [**], [***] Mean difference is significant at   = 0.05, 0.01, 0.001

Less familiar members of the public feel that making the permitting process easier is a more important objective. Interestingly, less familiar respondents are more supportive of increasing the number of acres in the National Forest System but are less likely to feel that this is an appropriate agency role. Here again, less familiar respondents give the USDA Forest Service a higher performance rating than do those with a higher degree of agency familiarity.

Strategic Plan Objective 2.d: Increase accessibility to a diversity of people and members of under served and low-income populations to the full range of uses, values, products, and services.

Increased accessibility for traditionally underserved populations is the fourth objective under the goal of providing benefits to people. Table 32 contains the NSRE results by USDA Forest Service familiarity.

Less familiar respondents see the easing of the permitting process as a slightly important objective, while those with the most agency familiarity see it as somewhat unimportant. This result needs to be qualified with the realization that respondents who are less familiar with the USDA Forest Service may also be unfamiliar with the permitting process. Less familiar respondents feel that the preservation of local cultural uses is a more important role for the agency than do those with more familiarity. Local decision making is seen as a more important agency function by low or moderately familiar respondents. And as before, attitudes about agency performance are most favorable among least familiar members of the public.

Table 32. Mean scores by respondent familiarity for Strategic Plan Objective 2.d.: Increase accessibility to a diversity of people and members of underserved and low-income populations to the full range of uses, values, products, and services. Survey items are from the VOBA module of the National Survey on Recreation and the Environment.

Survey item number and statement	Objectives for the management of forests and grasslands (1= not at all important, 5= very important)				
	Full sample N=5,064	Low familiarity N=942	Moderate familiarity N=3,533	High familiarity N=589	Signif. diff. Source of diff.
11. Preserve local cultural uses	3.82 1.38[a]	3.85 1.54	3.71 1.35	3.73 1.16	
12. Provide natural resources to dependent communities	3.60 1.39	3.67 1.39	3.58 1.39	3.31 1.24	
15. Make the permitting process easier for established uses	2.89 1.55	3.40 1.69	2.81 1.44	2.57 1.26	*** L-M.H[b]
16. Develop a national policy for natural resource development	4.26 1.23	4.28 1.31	4.27 1.12	4.33 1.05	
26. Make management decisions at the local level (rather than national)	3.93 1.22	4.00 1.30	3.87 1.14	4.18 0.99	

Survey item number and statement	Beliefs about the role of the USDA Forest Service (1= strongly disagree, 5= strongly agree)				
	Full sample N=5,064	Low familiarity N=942	Moderate familiarity N=3,533	High familiarity N=589	Signif. diff. Source of diff.
11. Preserve local cultural uses	3.75 1.36	3.89 1.52	3.74 1.31	3.50 1.10	* L-H
12. Provide natural resources to dependent communities	3.29 1.39	3.46 1.43	3.38 1.31	3.12 1.09	
15. Make the permitting process easier for established uses	2.90 1.59	2.70 1.81	2.77 1.43	2.45 1.40	
16. Develop a national policy for natural resource development	4.21 1.19	4.04 1.48	4.28 1.11	4.05 1.00	* L-M
26. Make management decisions at the local level (rather than national)	3.88 1.34	3.73 1.63	3.93 1.22	3.57 1.21	*

Survey item number and statement	Attitudes about the performance of the Forest Service (1= very unfavorable, 5= very favorable)				
	Full sample N=5,064	Low familiarity N=942	Moderate familiarity N=3,533	High familiarity N=589	Signif. diff. Source of diff.
11. Preserve local cultural uses	3.39 1.37	3.59 1.25	3.46 1.30	3.32 0.98	
12. Provide natural resources to dependent communities	3.45 1.30	3.21 1.53	3.54 1.16	2.90 1.09	*** M-L, H
15. Make the permitting process easier for established uses	3.05 1.43	3.08 1.75	3.10 1.34	2.60 0.97	
16. Develop a national policy for natural resource development	3.51 1.26	3.71 1.30	3.53 1.17	3.46 1.12	
26. Make management decisions at the local level (rather than national)	3.49 1.32	3.87 1.54	3.45 1.25	3.30 1.19	** L-M, H

[a] Standard deviation
[b] For example, L-M indicates that the source of the statistically significant difference lies in the differences between the responses of those with low agency familiarity and those with moderate familiarity
*, **, *** Mean difference is significant at = 0.05, 0.01, 0.001

USDA Forest Service RMRS GTR-95. 2002

Strategic Plan Objective 2.e: Improve delivery of services to urban communities.

The final objective supporting the goal of providing benefits to people is the improvement of delivery of services to urban communities. Table 33 shows the results for this objective by level of familiarity.

Expanded off-highway motorized access and the development of new paved roads and the expansion of commercial recreation are most supported by less familiar respondents. Those with moderate familiarity are most in favor of trails for non-motorized recreation. With this objective, we see again that those with lower agency familiarity see the preservation of a wilderness experience as a less important objective for public land management.

It is interesting that respondents most familiar with the agency do not see the delivery of services for metropolitan populations as an appropriate role for the USDA Forest Service. Where statistically significant differences exist, those with low or moderate agency familiarity see these items as more appropriate roles for the USDA Forest Service. Only slight differences in attitudes about agency performance exist, and those less familiar generally give higher ratings.

Table 33. Mean scores by respondent familiarity for Strategic Plan Objective 2.e.: Improve delivery of services to urban communities. Survey items are from the VOBA module of the National Survey on Recreation and the Environment.

Survey item number and statement	**Objectives** for the management of forests and grasslands (1= not at all important, 5= very important)				
	Full sample N=5,064	Low familiarity N=942	Moderate familiarity N=3,533	High familiarity N=589	Signif. diff. Source of diff.
1. Expand off-highway motorized access	2.41 *1.49[a]*	2.41 *1.86*	2.45 *1.40*	2.04 *1.13*	
2. Trails for motorized vehicles	2.82 *1.47*	2.98 *1.51*	2.83 *1.39*	2.99 *1.49*	
3. Trails for non-motorized recreation	3.75 *1.33*	3.49 *1.36*	3.82 *1.24*	3.50 *1.13*	** L-M[b]
4. Designate trails for specific uses	3.59 *1.43*	3.68 *1.58*	3.61 *1.37*	3.34 *1.16*	
5. Develop new paved roads	2.62 *1.49*	2.91 *1.75*	2.57 *1.39*	2.17 *1.14*	*** L-M, H
10. Preserve wilderness experience	4.15 *1.28*	3.99 *1.42*	4.16 *1.18*	3.96 *1.09*	
17. Expand commercial recreation	3.04 *1.45*	3.32 *1.76*	2.85 *1.25*	2.76 *1.12*	*** L-M, H

Survey item number and statement	**Beliefs** about the role of the USDA Forest Service (1= strongly disagree, 5= strongly agree)				
	Full sample N=5,064	Low familiarity N=942	Moderate familiarity N=3,533	High familiarity N=589	Signif. diff. Source of diff.
1. Expand off-highway motorized access	2.52 *1.50*	2.52 *1.60*	2.60 *1.43*	2.01 *1.07*	** H-L, M
2. Trails for motorized vehicles	2.98 *1.51*	3.20 *1.62*	2.92 *1.45*	2.65 *1.13*	* L-H
3. Trails for non-motorized recreation	3.71 *1.37*	3.77 *1.60*	3.67 *1.34*	3.47 *1.13*	

continued on next page

Table 33. *Continued.*

4. Designate trails for specific uses	3.94 *1.25*	3.88 *1.50*	4.00 *1.10*	3.57 *1.10*	* M-H
5. Develop new paved roads	2.70 *1.57*	2.89 *1.83*	2.57 *1.42*	1.92 *0.97*	*** L-M, H; M-H
10. Preserve wilderness experience	4.22 *1.14*	4.06 *1.29*	4.32 *1.05*	4.12 *1.08*	** L-M
17. Expand commercial recreation	3.25 *1.53*	3.13 *1.59*	3.24 *1.44*	2.78 *1.26*	* M-H

	Attitudes about the performance of the Forest Service *(1= very unfavorable, 5= very favorable)*				
Survey item number and statement	Full sample N=5,064	Low familiarity N=942	Moderate familiarity N=3,533	High familiarity N=589	Signif. diff. Source of diff.
1. Expand off-highway motorized access	2.97 *1.45*	3.16 *1.63*	2.90 *1.38*	3.10 *1.16*	
2. Trails for motorized vehicles	3.25 *1.38*	3.39 *1.48*	3.35 *1.28*	2.98 *1.01*	*
3. Trails for non-motorized recreation	3.59 *1.27*	3.47 *1.42*	3.73 *1.19*	3.24 *0.88*	* M-H
4. Designate trails for specific uses	3.61 *1.33*	3.44 *1.48*	3.71 *1.11*	3.39 *1.06*	* L-M
5. Develop new paved roads	3.19 *1.43*	3.33 *1.71*	3.18 *1.32*	3.01 *1.10*	
10. Preserve wilderness experience	3.88 *1.10*	3.80 *1.25*	3.93 *1.06*	3.72 *0.85*	
17. Expand commercial recreation	3.45 *1.24*	3.47 *1.37*	3.54 *1.21*	3.36 *0.99*	

[a] Standard deviation
[b] For example, L-M indicates that the source of the statistically significant difference lies in the differences between the responses of those with low agency familiarity and those with moderate familiarity
*, **, *** Mean difference is significant at = 0.05, 0.01, 0.001

Goal 3: Scientific and Technical Assistance: Develop and use the best scientific information available to deliver technical and community assistance and to support ecological, economic, and social sustainability.

The USDA Forest Service will develop and use the best scientific information available to deliver technical and community assistance and to support ecological, economic, and social sustainability.

Strategic Plan Objective 3.a: Better assist in building the capacity of Tribal governments, rural communities, and private landowners to adapt to economic, environmental, and social change related to natural resources.

The first objective supporting scientific and technical assistance deals with increasing the capacity of rural communities, tribal governments, and private landowners to deal with changing USDA Forest Service management. Table 34 shows the results for this objective by familiarity with the USDA Forest Service.

Less familiar members of the public find the provision of information on recreation concerns to be less important than do those more familiar with the agency. Those with moderate familiarity are most supportive of encouraging collaboration among groups. This group also feels that the encouragement of collaboration is a less important role for the agency.

Table 34. Mean scores by respondent familiarity for Strategic Plan Objective 3.a.: Better assist in building the capacity of Tribal governments, rural communities, and private landowners to adapt to economic, environmental, and social change related to natural resources. Survey items are from the VOBA module of the National Survey on Recreation and the Environment.

| Survey item number and statement | **Objectives** for the management of forests and grasslands *(1= not at all important, 5= very important)* | | | | |
	Full sample N=5,064	Low familiarity N=942	Moderate familiarity N=3,533	High familiarity N=589	Signif. diff. Source of diff.
11. Preserve local cultural uses	3.82 *1.38[a]*	3.85 *1.54*	3.71 *1.35*	3.73 *1.16*	
20. Inform the public about recreation concerns (safety, trail etiquette, etc.)	4.53 *0.93*	4.37 *1.18*	4.57 *0.84*	4.63 *0.63*	* L-M[b]
21. Inform the public on potential environmental impacts of uses	4.39 *1.02*	4.30 *1.13*	4.48 *0.90*	4.44 *0.81*	
22. Inform the public on economic value from developing natural resources	4.02 *1.30*	4.18 *1.34*	4.03 *1.20*	3.95 *1.13*	
23. Encourage collaboration between groups to share information	4.15 *1.20*	3.97 *1.46*	4.27 *1.05*	3.81 *1.10*	*** M-L, H; L-H
24. Use public advisory committees to advise on management issues	3.90 *1.20*	3.94 *1.18*	3.86 *1.17*	3.77 *1.09*	
26. Make management decisions at the local level (rather than national)	3.93 *1.22*	4.00 *1.30*	3.87 *1.14*	4.18 *0.99*	

| Survey item number and statement | **Beliefs** about the role of the USDA Forest Service *(1= strongly disagree, 5= strongly agree)* | | | | |
	Full sample N=5,064	Low familiarity N=942	Moderate familiarity N=3,533	High familiarity N=589	Signif. diff. Source of diff.
11. Preserve local cultural uses	3.75 *1.36*	3.89 *1.52*	3.74 *1.31*	3.50 *1.10*	* L-H
20. Inform the public about recreation concerns (safety, trail etiquette, etc.)	4.50 *0.93*	4.57 *0.91*	4.53 *0.91*	4.47 *0.61*	
21. Inform the public on potential environmental impacts of uses	4.48 *1.00*	4.46 *1.07*	4.46 *0.94*	4.13 *1.00*	* H-L, M
22. Inform the public on economic value from developing natural resources	4.08 *1.17*	3.91 *1.37*	4.12 *1.12*	3.76 *1.06*	* M-H
23. Encourage collaboration between groups to share information	4.15 *1.17*	3.84 *1.48*	4.20 *1.03*	4.42 *0.76*	*** L-M, H
24. Use public advisory committees to advise on management issues	3.84 *1.25*	3.75 *1.41*	3.83 *1.19*	3.84 *1.14*	
26. Make management decisions at the local level (rather than national)	3.88 *1.34*	3.73 *1.63*	3.93 *1.22*	3.57 *1.21*	*

| Survey item number and statement | **Attitudes** about the performance of the Forest Service *(1= very unfavorable, 5= very favorable)* | | | | |
	Full sample N=5,064	Low familiarity N=942	Moderate familiarity N=3,533	High familiarity N=589	Signif. diff. Source of diff.
11. Preserve local cultural uses	3.39 *1.37*	3.59 *1.25*	3.46 *1.30*	3.32 *0.98*	

continued on next page

Table 34. *Continued.*

	Full sample	Low familiarity	Moderate familiarity	High familiarity	Signif. diff.
20. Inform the public about recreation concerns (safety, trail etiquette, etc.)	3.89 *1.31*	3.95 *1.48*	3.89 *1.21*	3.83 *1.05*	
21. Inform the public on potential environmental impacts of uses	3.50 *1.33*	3.66 *1.45*	3.49 *1.23*	3.02 *1.08*	** H-L, M
22. Inform the public on economic value from developing natural resources	3.40 *1.38*	3.51 *1.38*	3.43 *1.32*	2.63 *1.08*	*** H-L, M
23. Encourage collaboration between groups to share information	3.72 *1.22*	3.75 *1.25*	3.75 *1.15*	3.52 *1.07*	
24. Use public advisory committees to advise on management issues	3.36 *1.26*	3.68 *1.14*	3.38 *1.19*	3.06 *0.91*	* L-H
26. Make management decisions at the local level (rather than national)	3.49 *1.32*	3.87 *1.54*	3.45 *1.25*	3.30 *1.19*	** L-M, H

[a] Standard deviation
[b] For example, L-M indicates that the source of the statistically significant difference lies in the differences between the responses of those with low agency familiarity and those with moderate familiarity
*, **, *** Mean difference is significant at α = 0.05, 0.01, 0.001

Less familiar respondents see the provision of information on potential environmental impacts as a more important objective for public lands, and they also rate the agency performance in this area higher than those more familiar. Other items with statistically different attitudes on agency performance are again rated higher by less familiar respondents.

Strategic Plan Objective 3.b: Increase the effectiveness of scientific, developmental, and technical information delivered to domestic and international interests.

The second Strategic Plan objective supporting the provision of scientific and technical assistance deals with increasing the effectiveness of such assistance. The VOBA items dealing with this objective, presented by level of agency familiarity, are in table 35.

Support for informing the public about recreation concerns increases as familiarity with the USDA Forest Service increases. As respondents are more familiar with the agency, however, they are less likely to feel that it is an appropriate role for the USDA Forest Service to

Table 35. Mean scores by respondent familiarity for Strategic Plan Objective 3.b.: Increase the effectiveness of scientific, developmental, and technical information delivered to domestic and international interests. Survey items are from the VOBA module of the National Survey on Recreation and the Environment.

	Objectives for the management of forests and grasslands (1= not at all important, 5= very important)				
Survey item number and statement	Full sample N=5,064	Low familiarity N=942	Moderate familiarity N=3,533	High familiarity N=589	Signif. diff. Source of diff.
20. Inform the public about recreation concerns (safety, trail etiquette, etc.)	4.53 *0.93*[a]	4.37 *1.18*	4.57 *0.84*	4.63 *0.63*	* L-M[b]
21. Inform the public on potential environmental impacts of uses	4.39 *1.02*	4.30 *1.13*	4.48 *0.90*	4.44 *0.81*	
22. Inform the public on economic value from developing natural resources	4.02 *1.30*	4.18 *1.34*	4.03 *1.20*	3.95 *1.13*	
23. Encourage collaboration between groups to share information	4.15 *1.20*	3.97 *1.46*	4.27 *1.05*	3.81 *1.10*	*** M-L, H

continued on next page

Table 35. *Continued.*

	Beliefs about the role of the USDA Forest Service *(1= strongly disagree, 5= strongly agree)*				
Survey item number and statement	Full sample N=5,064	Low familiarity N=942	Moderate familiarity N=3,533	High familiarity N=589	Signif. diff. Source of diff.
20. Inform the public about recreation concerns (safety, trail etiquette, etc.)	4.50 *0.93*	4.57 *0.91*	4.53 *0.91*	4.47 *0.61*	
21. Inform the public on potential environmental impacts of uses	4.48 *1.00*	4.46 *1.07*	4.46 *0.94*	4.13 *1.00*	* H-L, M
22. Inform the public on economic value from developing natural resources	4.08 *1.17*	3.91 *1.37*	4.12 *1.12*	3.76 *1.06*	* M-H
23. Encourage collaboration between groups to share information	4.15 *1.17*	3.84 *1.48*	4.20 *1.03*	4.42 *0.76*	*** L-M, H

	Attitudes about the performance of the Forest Service *(1= very unfavorable, 5= very favorable)*				
Survey item number and statement	Full sample N=5,064	Low familiarity N=942	Moderate familiarity N=3,533	High familiarity N=589	Signif. diff. Source of diff.
20. Inform the public about recreation concerns (safety, trail etiquette, etc.)	3.89 *1.31*	3.95 *1.48*	3.89 *1.21*	3.83 *1.05*	
21. Inform the public on potential environmental impacts of uses	3.50 *1.33*	3.66 *1.45*	3.49 *1.23*	3.02 *1.08*	** H-L, M
22. Inform the public on economic value from developing natural resources	3.40 *1.38*	3.51 *1.38*	3.43 *1.32*	2.63 *1.08*	*** H-L, M
23. Encourage collaboration between groups to share information	3.72 *1.22*	3.75 *1.25*	3.75 *1.15*	3.52 *1.07*	

[a] Standard deviation
[b] For example, L-M indicates that the source of the statistically significant difference lies in the differences between the responses of those with low agency familiarity and those with moderate familiarity
*, **, *** Mean difference is significant at = 0.05, 0.01, 0.001

provide information about the potential environmental impacts of activities. Those with a moderate level of familiarity are most supportive of the agency's role in informing the public about economic benefits from natural resources. Respondents with a high degree of familiarity see encouraging collaboration as a more important role for the agency. Low familiarity corresponds with higher agency performance ratings where statistical differences occur.

Strategic Plan Objective 3.c: Improve the knowledge base provided through research, inventory, and monitoring to enhance scientific understanding of ecosystems, including human uses, and to support decision-making and sustainable management of the Nation's forests and grasslands.

The provision of scientific and technical assistance includes research efforts to improve the understanding of ecosystems, as described in the third Strategic Plan objective under this goal. VOBA items for this objective by familiarity are in table 36.

Support for informing the public about recreation concerns increases as familiarity with the USDA Forest Service increases. As respondents are more familiar with the agency, however, they are less likely to feel that it is an appropriate role for the USDA Forest Service

Table 36. Mean scores by respondent familiarity for Strategic Plan Objective 3.c.: Improve the knowledge base provided through research, inventory, and monitoring to enhance scientific understanding of ecosystems, including human uses, and to support decision-making and sustainable management of the Nation's forests and grasslands. Survey items are from the VOBA module of the National Survey on Recreation and the Environment.

| Survey item number and statement | **Objectives** for the management of forests and grasslands (1= not at all important, 5= very important) | | | | |
	Full sample N=5,064	Low familiarity N=942	Moderate familiarity N=3,533	High familiarity N=589	Signif. diff. Source of diff.
20. Inform the public about recreation concerns (safety, trail etiquette, etc.)	4.53 *0.93[a]*	4.37 *1.18*	4.57 *0.84*	4.63 *0.63*	* L-M[b]
21. Inform the public on potential environmental impacts of uses	4.39 *1.02*	4.30 *1.13*	4.48 *0.90*	4.44 *0.81*	
22. Inform the public on economic value from developing natural resources	4.02 *1.30*	4.18 *1.34*	4.03 *1.20*	3.95 *1.13*	
23. Encourage collaboration between groups to share information	4.15 *1.20*	3.97 *1.46*	4.27 *1.05*	3.81 *1.10*	*** M-L, H

| Survey item number and statement | **Beliefs** about the role of the USDA Forest Service (1= strongly disagree, 5= strongly agree) | | | | |
	Full sample N=5,064	Low familiarity N=942	Moderate familiarity N=3,533	High familiarity N=589	Signif. diff. Source of diff.
20. Inform the public about recreation concerns (safety, trail etiquette, etc.)	4.50 *0.93*	4.57 *0.91*	4.53 *0.91*	4.47 *0.61*	
21. Inform the public on potential environmental impacts of uses	4.48 *1.00*	4.46 *1.07*	4.46 *0.94*	4.13 *1.00*	* H-L, M
22. Inform the public on economic value from developing natural resources	4.08 *1.17*	3.91 *1.37*	4.12 *1.12*	3.76 *1.06*	* M-H
23. Encourage collaboration between groups to share information	4.15 *1.17*	3.84 *1.48*	4.20 *1.03*	4.42 *0.76*	*** L-M, H

| Survey item number and statement | **Attitudes** about the performance of the Forest Service (1= very unfavorable, 5= very favorable) | | | | |
	Full sample N=5,064	Low familiarity N=942	Moderate familiarity N=3,533	High familiarity N=589	Signif. diff. Source of diff.
20. Inform the public about recreation concerns (safety, trail etiquette, etc.)	3.89 *1.31*	3.95 *1.48*	3.89 *1.21*	3.83 *1.05*	
21. Inform the public on potential environmental impacts of uses	3.50 *1.33*	3.66 *1.45*	3.49 *1.23*	3.02 *1.08*	** H-L, M
22. Inform the public on economic value from developing natural resources	3.40 *1.38*	3.51 *1.38*	3.43 *1.32*	2.63 *1.08*	*** H-L, M
23. Encourage collaboration between groups to share information	3.72 *1.22*	3.75 *1.25*	3.75 *1.15*	3.52 *1.07*	

[a] Standard deviation
[b] For example, L-M indicates that the source of the statistically significant difference lies in the differences between the responses of those with low agency familiarity and those with moderate familiarity
*, **, *** Mean difference is significant at = 0.05, 0.01, 0.001

to provide information about the potential environmental impacts of activities. Those with a moderate level of familiarity are most supportive of the agency's role in informing the public about economic benefits from natural resources. Respondents with a high degree of familiarity see encouraging collaboration as a more important role for the agency. Low familiarity corresponds with higher agency performance ratings where statistical differences occur.

Strategic Plan Objective 3.d: Broaden the participation of less traditional research groups in research and technical assistance programs.

The final objective under the goal of providing scientific and technical assistance is to broaden participation in research and technical assistance programs. Table 37 shows the VOBA results by familiarity for this Strategic Plan objective.

As an objective for public lands, the encouragement of collaboration is least supported by those with the highest level of agency familiarity, but this same group sees this activity as a more important role for the agency than those less familiar.

Table 37. Mean scores by respondent familiarity for Strategic Plan Objective 3.d.: Broaden the participation of less traditional research groups in research and technical assistance programs. Survey items are from the VOBA module of the National Survey on Recreation and the Environment.

	Objectives for the management of forests and grasslands (*1= not at all important, 5= very important*)				
Survey item number and statement	Full sample N=5,064	Low familiarity N=942	Moderate familiarity N=3,533	High familiarity N=589	Signif. diff. Source of diff.
23. Encourage collaboration between groups to share information	4.15 *1.20*[a]	3.97 *1.46*	4.27 *1.05*	3.81 *1.10*	*** M-L, H[b]
	Beliefs about the role of the USDA Forest Service (*1= strongly disagree, 5= strongly agree*)				
Survey item number and statement	Full sample N=5,064	Low familiarity N=942	Moderate familiarity N=3,533	High familiarity N=589	Signif. diff. Source of diff.
23. Encourage collaboration between groups to share information	4.15 *1.17*	3.84 *1.48*	4.20 *1.03*	4.42 *0.76*	*** L-M, H
	Attitudes about the performance of the Forest Service (*1= very unfavorable, 5= very favorable*)				
Survey item number and statement	Full sample N=5,064	Low familiarity N=942	Moderate familiarity N=3,533	High familiarity N=589	Signif. diff. Source of diff.
23. Encourage collaboration between groups to share information	3.72 *1.22*	3.75 *1.25*	3.75 *1.15*	3.52 *1.07*	

[a] Standard deviation

[b] For example, L-M indicates that the source of the statistically significant difference lies in the differences between the responses of those with low agency familiarity and those with moderate familiarity

*, **, *** Mean difference is significant at = 0.05, 0.01, 0.001

Goal 4: Effective Public Service: Ensure the acquisition and use of an appropriate corporate infrastructure to enable the efficient delivery of a variety of uses.

The USDA Forest Service will ensure the acquisition and use of an appropriate corporate structure to enable the efficient delivery of a variety of uses. To meet the goal of effective public service the USDA Forest Service has described six objectives. Three of these can be addressed using the VOBA results.

Strategic Plan Objective 4.a: Improve financial management to achieve fiscal accountability.

The first objective is to improve financial management to achieve fiscal accountability. See table 38 for VOBA results by agency familiarity.

While no group feels that subsidizing commodity development is appropriate, those who are most familiar with the USDA Forest Service are least in favor of such subsidies.

Table 38. Mean scores by respondent familiarity for Strategic Plan Objective 4.a.: Improve financial management to achieve fiscal accountability. Survey item is from the VOBA module of the National Survey on Recreation and the Environment.

	Values				
	with respect to forests and grasslands				
	(1= strongly disagree, 5=strongly agree)				
Survey item number and statement	Full sample N=5,064	Low familiarity N=942	Moderate familiarity N=3,533	High familiarity N=589	Signif. diff. Source of diff.
24. The federal government should subsidize development and leasing of public lands	2.32 *1.58*[a]	2.46 *1.88*	2.19 *1.42*	1.77 *1.09*	*** L-M, H; M-H[b]

[a] Standard deviation
[b] For example, L-M indicates that the source of the statistically significant difference lies in the differences between the responses of those with low agency familiarity and those with moderate familiarity
*, **, *** Mean difference is significant at = 0.05, 0.01, 0.001

Strategic Plan Objective 4.b: Improve the safety and economy of USDA Forest Service roads, trails, facilities, and operations and provide greater security for the public and employees.

A second objective supporting the improvement of public service is to increase the safety and economy of USDA Forest Service facilities. Table 39 shows the VOBA results by agency familiarity.

Less familiar respondents are less in favor of informing the public about recreation concerns on public lands, but are more in favor of increased law enforcement. This group also sees increasing law enforcement as a more appropriate role for the USDA Forest Service. As has been the case for most of the attitudes items, where statistically significant differences exist, those with lower levels of agency familiarity give the agency a higher rating.

Table 39. Mean scores by respondent familiarity for Strategic Plan Objective 4.b.: Improve the safety an economy of USDA Forest Service roads, trails, facilities, and operations an provide greater security for the public and employees. Survey items are from the VOBA module of the National Survey on Recreation and the Environment.

Survey item number and statement	VOBA Objectives for the management of forests and grasslands (1= not at all important, 5= very important)				
	Full sample N=5,064	Low familiarity N=942	Moderate familiarity N=3,533	High familiarity N=589	Signif. diff. Source of diff.
19. Develop volunteer programs to improve facilities (trails, etc.)	4.18 *1.13[a]*	4.09 *1.47*	4.19 *1.02*	4.20 *0.88*	
20. Inform the public about recreation concerns (safety, trail etiquette, etc.)	4.53 *0.93*	4.37 *1.18*	4.57 *0.84*	4.63 *0.63*	* L-M[b]
29. Increase law enforcement on National Forests and Grasslands	4.01 *1.21*	4.16 *1.18*	3.88 *1.22*	3.77 *0.89*	* L-M

Survey item number and statement	VOBA Beliefs about the role of the USDA Forest Service (1= strongly disagree, 5= strongly agree)				
	Full sample N=5,064	Low familiarity N=942	Moderate familiarity N=3,533	High familiarity N=589	Signif. diff. Source of diff.
19. Develop volunteer programs to improve facilities (trails, etc.)	4.22 *1.09*	4.24 *1.15*	4.28 *0.99*	4.14 *0.85*	
20. Inform the public about recreation concerns (safety, trail etiquette, etc.)	4.50 *0.93*	4.57 *0.91*	4.53 *0.91*	4.47 *0.61*	
29. Increase law enforcement on National Forests and Grasslands	4.01 *1.26*	4.12 *1.30*	4.02 *1.17*	3.47 *1.29*	** H-L, M

Survey item number and statement	VOBA Attitudes about the performance of the Forest Service (1= very unfavorable, 5= very favorable)				
	Full sample N=5,064	Low familiarity N=942	Moderate familiarity N=3,533	High familiarity N=589	Signif. diff. Source of diff.
19. Develop volunteer programs to improve facilities (trails, etc.)	3.79 *1.21*	4.09 *1.14*	3.82 *1.09*	3.57 *1.02*	** L-M, H
20. Inform the public about recreation concerns (safety, trail etiquette, etc.)	3.89 *1.31*	3.95 *1.48*	3.89 *1.21*	3.83 *1.05*	
29. Increase law enforcement on National Forests and Grasslands	3.85 *1.27*	4.10 *1.31*	3.91 *1.18*	3.48 *1.08*	** H-L, M

[a] Standard deviation
[b] For example, L-M indicates that the source of the statistically significant difference lies in the differences between the responses of those with low agency familiarity and those with moderate familiarity
*, **, *** Mean difference is significant at = 0.05, 0.01, 0.001

Strategic Plan Objective 4.f: Provide appropriate access to National Forest System lands and ensure nondiscrimination in the delivery of all USDA Forest Service programs.

Providing access is the final objective under the goal of improving public service. VOBA results for this objective are in table 40.

Higher support for easing the permitting process and for expanding commercial recreation exists among those with less familiarity with the USDA Forest Service. These respondents also see the expansion of commercial recreation as a more important role for the agency.

Table 40. Mean scores by respondent familiarity for Strategic Plan Objective 4.f.: Provide appropriate access to National Forest System lands and ensure nondiscrimination in the delivery of all USDA Forest Service programs. Survey items are from the VOBA module of the National Survey on Recreation and the Environment.

Survey item number and statement	**Objectives** for the management of forests and grasslands (*1= not at all important, 5= very important*)				
	Full sample N=5,064	Low familiarity N=942	Moderate familiarity N=3,533	High familiarity N=589	Signif. diff. Source of diff.
11. Preserve local cultural uses	3.82 *1.38*[a]	3.85 *1.54*	3.71 *1.35*	3.73 *1.16*	
12. Provide natural resources to dependent communities	3.60 *1.39*	3.67 *1.39*	3.58 *1.39*	3.31 *1.24*	
15. Make the permitting process easier for established uses	2.89 *1.55*	3.40 *1.69*	2.81 *1.44*	2.57 *1.26*	*** L-M, H[b]
17. Expand commercial recreation	3.04 *1.45*	3.32 *1.76*	2.85 *1.25*	2.76 *1.12*	*** L-M, H
25. Allow for diverse uses	4.07 *1.18*	3.94 *1.42*	4.15 *1.05*	3.91 *0.91*	*

Survey item number and statement	**Beliefs** about the role of the USDA Forest Service (*1= strongly disagree, 5= strongly agree*)				
	Full sample N=5,064	Low familiarity N=942	Moderate familiarity N=3,533	High familiarity N=589	Signif. diff. Source of diff.
11. Preserve local cultural uses	3.75 *1.36*	3.89 *1.52*	3.74 *1.31*	3.50 *1.10*	* L-H
12. Provide natural resources to dependent communities	3.29 *1.39*	3.46 *1.43*	3.38 *1.31*	3.12 *1.09*	
15. Make the permitting process easier for established uses	2.90 *1.59*	2.70 *1.81*	2.77 *1.43*	2.45 *1.40*	
17. Expand commercial recreation	3.25 *1.53*	3.13 *1.59*	3.24 *1.44*	2.78 *1.26*	* M-H
25. Allow for diverse uses	4.02 *1.14*	3.96 *1.18*	4.04 *1.12*	4.13 *0.90*	

Survey item number and statement	**Attitudes** about the performance of the Forest Service (*1= very unfavorable, 5= very favorable*)				
	Full sample N=5,064	Low familiarity N=942	Moderate familiarity N=3,533	High familiarity N=589	Signif. diff. Source of diff.
11. Preserve local cultural uses	3.39 *1.37*	3.59 *1.25*	3.46 *1.30*	3.32 *0.98*	
12. Provide natural resources to dependent communities	3.45 *1.30*	3.21 *1.53*	3.54 *1.16*	2.90 *1.09*	*** M-L, H
15. Make the permitting process easier for established uses	3.05 *1.43*	3.08 *1.75*	3.10 *1.34*	2.60 *0.97*	
17. Expand commercial recreation	3.45 *1.24*	3.47 *1.37*	3.54 *1.21*	3.36 *0.99*	
25. Allow for diverse uses	3.73 *1.18*	3.91 *1.10*	3.82 *1.10*	3.44 *0.98*	* H-L, M

[a] Standard deviation
[b] For example, L-M indicates that the source of the statistically significant difference lies in the differences between the responses of those with low agency familiarity and those with moderate familiarity
*, **, *** Mean difference is significant at = 0.05, 0.01, 0.001

Those with a moderate level of agency familiarity are most in favor of allowing diverse uses of public lands.

The preservation of local cultural uses is most supported as an agency role by the less familiar respondents. These respondents also give the agency higher performance ratings.

2. Survey Results for the Public Lands Values by Agency Familiarity

While no clear pattern emerges for the Socially Responsible Individual Values with respect to agency familiarity, one sees that those more familiar with the USDA Forest Service are more likely to disagree with the eight values statements that comprise the Socially Responsible Management Values. The items in this factor are worded such that a lower number (a response in "disagree" end of the scale) indicates a more "environmentally" oriented perspective. See table 41.

Table 41. Mean scores by respondent familiarity for public lands values. Survey items are from the VOBA module of the National Survey on Recreation and the Environment.

Survey item number and statement	Group A. Socially Responsible Individual Values [a]				
	Full sample N=5,064	Low familiarity N=942	Moderate familiarity N=3,533	High familiarity N=589	Signif. diff. Source of diff.
1. People should be more concerned about how our public lands are used	4.73 0.74[b]	4.80 0.79	4.72 0.75	4.64 0.59	
2. Natural resources must be preserved even if people must do without some products	4.12 1.20	4.25 1.22	4.12 1.20	3.82 1.17	* L-H[c]
3. Consumers should be interest in the environmental consequences of the products they purchase	4.46 0.96	4.33 1.30	4.52 0.87	4.35 0.75	* L-M
4. I would be willing to sign a petition for an environmental cause	4.00 1.34	4.13 1.40	3.95 1.35	3.96 1.17	
5r.[d] The whole pollution issue has upset me, I feel it's not overrated	3.70 1.45	3.34 1.73	3.82 1.38	3.83 1.31	*** L-M, L-H
6. I have often thought that if we could just get by with a little less there would be more left for future generations	4.00 1.35	3.97 1.64	4.04 1.28	3.82 1.32	
7. Manufacturers should be encourage to use recycled materials in their operations	4.65 0.84	4.58 1.06	4.66 0.79	4.69 0.69	
8. Future generations should be as important as the current one in decisions about public lands	4.54 0.96	4.57 1.05	4.52 0.96	4.57 0.78	
9. I would be willing to pay five dollars more each time I use public lands for recreation	3.50 1.57	3.53 1.87	3.51 1.53	3.38 1.31	
10. People should urge their friends to limit their use of products made from scarce resources	4.16 1.14	4.12 1.23	4.22 1.10	3.86 1.19	* M-H
11. I am glad there are national forests even if I never get to see them	4.72 0.80	4.63 0.97	4.74 0.75	4.71 0.78	
12. People can think public lands are valuable even if they do not actually go there themselves	4.59 0.88	4.52 1.16	4.66 0.76	4.32 0.93	** M-H
13. I am willing to stop buying products from companies that pollute	3.89 1.32	3.74 1.64	3.95 1.23	3.90 1.25	

continued on next page

Table 41. *Continued.*

	Full sample	Low familiarity	Moderate familiarity	High familiarity	Signif. diff.
14. I am willing to make personal sacrifices for the sake of slowing down pollution	4.31 *1.09*	4.17 *1.32*	4.38 *1.04*	4.26 *0.93*	* L-M
15. Forests have right to exist for their own sake	4.13 *1.23*	4.15 *1.36*	4.15 *1.20*	3.95 *1.25*	
16. Wildlife, plants and humans have equal rights to live and grow	4.27 *1.24*	4.55 *1.16*	4.20 *1.25*	4.05 *1.24*	** L-M, L-H
17. Donating time or money to worthy causes is important to me	4.22 *1.03*	4.25 *1.10*	4.21 *1.03*	4.16 *0.94*	

Survey item number and statement	**Group B. Socially Responsible Management Values**				
	Full sample N=5,064	Low familiarity N=942	Moderate familiarity N=3,533	High familiarity N=589	Signif. diff. Source of diff.
18. We should actively harvest more trees for a much larger human population	2.83 *1.66*	3.30 *1.97*	2.73 *1.59*	2.28 *1.22*	*** L-M, L-H, M-H
19. The most important role for public lands is providing jobs and income for local people	3.16 *1.46*	3.63 *1.68*	3.08 *1.39*	2.31 *1.04*	*** L-M, L-H, M-H
20. The decision to develop resources should be made mostly on economic grounds	2.94 *1.46*	3.23 *1.53*	2.95 *1.47*	2.04 *0.99*	*** L-M, L-H, M-H
21. The main reason for maintaining resources today is so we can develop them in the future	4.01 *1.34*	4.34 *1.30*	3.98 *1.31*	3.42 *1.37*	*** L-M, L-H, M-H
22. I think the public land managers are doing an adequate job of protecting	3.20 *1.27*	3.27 *1.47*	3.23 *1.23*	2.79 *1.07*	** L-H, M-H
23. The primary use of forests should be for products that are useful to humans	2.92 *1.59*	3.34 *1.80*	2.87 *1.57*	2.28 *1.14*	*** L-M, L-H, M-H
24. The federal government should subsidize the development and leasing of public lands	2.22 *1.48*	2.46 *1.88*	2.19 *1.42*	1.77 *1.09*	*** L-M, L-H, M-H
25. The government has better places to spend money than a strong conservation program	2.30 *1.34*	2.51 *1.51*	2.27 *1.32*	1.90 *1.02*	*** L-M, L-H, M-H
Group mean	**2.95**	**3.26**	**2.91**	**2.35**	

[a] 1=strongly disagree, 5=strongly agree
[b] Standard deviation
[c] For example, L-M indicates that the source of the statistically significant difference lies in the differences between the responses of those with low agency familiarity and those with moderate familiarity
*, **, *** Mean difference is significant at = 0.05, 0.01, 0.001
[d] Values statement 5 has been reverse scored in order to calculate a group mean. See Appendix A for an explanation of reverse scoring

3. Survey Objectives, Beliefs, and Attitudes Grouped by Strategic Level Objectives, by Familiarity With the USDA Forest Service

Access

Respondents with low or moderate agency familiarity offer stronger support for the expansion of off-highway motorized access, trails for non-motorized recreation, and the development of more paved roads. These respondents also see the agency as having a stronger role in the expansion of off-highway motorized access, in providing trails for motorized vehicles, in designating trails for specific uses, and in development of paved roads.

Those most familiar with the agency rate the performance in the area of expanding off-highway motorized access higher than other groups, while those with low or moderate familiarity are more likely to approve of USDA Forest Service performance in other areas related to access. See table 42.

Table 42. Mean scores by respondent familiarity for access. Survey items are from the VOBA module of the National Survey on Recreation and the Environment.

Survey item number and statement	Objectives for the management of forests and grasslands (1= not at all important, 5= very important)				
	Full sample N=5,064	Low familiarity N=942	Moderate familiarity N=3,533	High familiarity N=589	Signif. diff. Source of diff.
1. Expand off-highway motorized access	2.41 _1.49[a]_	2.41 _1.86_	2.45 _1.40_	2.04 _1.13_	
2. Increase trails for motorized vehicles	2.82 _1.47_	2.98 _1.51_	2.83 _1.39_	2.99 _1.49_	
3. Increase trails for non-motorized recreation	3.75 _1.33_	3.49 _1.36_	3.82 _1.24_	3.50 _1.13_	** L-M[b]
4. Designate trails for specific uses	3.59 _1.43_	3.68 _1.58_	3.61 _1.37_	3.34 _1.16_	
5. Develop more paved roads	2.62 _1.49_	2.91 _1.75_	2.57 _1.39_	2.17 _1.14_	*** L-M, H
6. Designate more wilderness to stop development & motorized access	3.84 _1.41_	3.65 _1.67_	3.83 _1.30_	3.99 _1.14_	

Survey item number and statement	Beliefs about the role of the USDA Forest Service (1= strongly disagree, 5= strongly agree)				
	Full sample N=5,064	Low familiarity N=942	Moderate familiarity N=3,533	High familiarity N=589	Signif. diff. Source of diff.
1. Expand off-highway motorized access	2.52 _1.50_	2.52 _1.60_	2.60 _1.43_	2.01 _1.07_	** H-L, M
2. Increase trails for motorized vehicles	2.98 _1.51_	3.20 _1.62_	2.92 _1.45_	2.65 _1.13_	* L-H
3. Increase trails for non-motorized recreation	3.71 _1.37_	3.77 _1.60_	3.67 _1.34_	3.47 _1.13_	
4. Designate trails for specific uses	3.94 _1.25_	3.88 _1.50_	4.00 _1.10_	3.57 _1.10_	* M-H
5. Develop more paved roads	2.70 _1.57_	2.89 _1.83_	2.57 _1.42_	1.92 _0.97_	*** L-M, H; M-H
6. Designate more wilderness to stop development & motorized access	3.66 _1.46_	3.57 _1.50_	3.71 _1.39_	3.61 _1.39_	

Survey item number and statement	Attitudes about the performance of the Forest Service (1= very unfavorable, 5= very favorable)				
	Full sample N=5,064	Low familiarity N=942	Moderate familiarity N=3,533	High familiarity N=589	Signif. diff. Source of diff.
1. Expand off-highway motorized access	3.25 _1.38_	3.39 _1.48_	3.35 _1.28_	2.98 _1.01_	*
2. Increase trails for motorized vehicles	3.25 _1.38_	3.35 _1.79_	3.31 _1.31_	2.99 _1.03_	
3. Increase trails for non-motorized recreation	3.59 _1.27_	3.47 _1.42_	3.73 _1.19_	3.24 _0.88_	* M-H
4. Designate trails for specific uses	3.61 _1.33_	3.44 _1.48_	3.71 _1.11_	3.39 _1.06_	* L-M
5. Develop more paved roads	3.19 _1.43_	3.33 _1.71_	3.18 _1.32_	3.01 _1.10_	
6. Designate more wilderness to stop	3.45 _1.34_	3.48 _1.35_	3.30 _1.30_	3.16 _1.21_	

[a] Standard deviation
[b] For example, L-M indicates that the source of the statistically significant difference lies in the differences between the responses of those with low agency familiarity and those with moderate familiarity
*, **, *** Mean difference is significant at = 0.05, 0.01, 0.001

Preservation/Conservation

The next set of items are related to the strategic level objective of preserving or conserving natural resources. Categorizations by agency familiarity are in table 43 for this objective.

Highly familiar members of the public are less likely to support policies that eliminate timber harvest and mining in order to preserve natural resources. These same respondents are also less likely to see the protection of ecosystems and watershed and the preservation of a wilderness experience as appropriate roles for the USDA Forest Service.

People more familiar with the agency also see the performance of the USDA Forest Service in the areas of watershed and ecosystem and habitat protection as less favorable than those less familiar.

Table 43. Mean scores by respondent familiarity for preservation/conservation. Survey items are from the VOBA module of the National Survey on Recreation and the Environment.

Survey item number and statement	**Objectives** for the management of forests and grasslands (1= not at all important, 5= very important)				
	Full sample N=5,064	Low familiarity N=942	Moderate familiarity N=3,533	High familiarity N=589	Signif. diff. Source of diff.
7. Conserve and protect watersheds	4.73 *0.76*[a]	4.70 *0.87*	4.72 *0.71*	4.74 *0.68*	
8. Preserve natural resources through policies such as no timber, no mining	4.22 *1.23*	4.18 *1.46*	4.31 *1.07*	3.99 *1.14*	* M-H[b]
9. Protecting ecosystems and wildlife habitat	4.58 *0.92*	4.48 *1.19*	4.59 *0.84*	4.55 *0.88*	
10. Preserve wilderness experience	4.15 *1.28*	3.99 *1.42*	4.16 *1.18*	3.96 *1.09*	
11. Preserve local cultural uses	3.82 *1.38*	3.85 *1.54*	3.71 *1.35*	3.73 *1.16*	

Survey item number and statement	**Beliefs** about the role of the USDA Forest Service (1= strongly disagree, 5= strongly agree)				
	Full sample N=5,064	Low familiarity N=942	Moderate familiarity N=3,533	High familiarity N=589	Signif. diff. Source of diff.
7. Conserve and protect watersheds	4.61 *0.83*	4.60 *0.88*	4.62 *0.79*	4.61 *0.64*	
8. Preserve natural resources through policies such as no timber, no mining	4.21 *1.27*	4.23 *1.31*	4.22 *1.09*	3.98 *1.22*	
9. Protecting ecosystems and wildlife habitat	4.53 *0.98*	4.40 *1.26*	4.55 *0.91*	4.23 *1.24*	* M-H
10. Preserve wilderness experience	4.22 *1.14*	4.06 *1.29*	4.32 *1.05*	4.12 *1.08*	** L-M
11. Preserve local cultural uses	3.75 *1.36*	3.89 *1.52*	3.74 *1.31*	3.50 *1.10*	* L-H

Survey item number and statement	**Attitudes** about the performance of the Forest Service (1= very unfavorable, 5= very favorable)				
	Full sample N=5,064	Low familiarity N=942	Moderate familiarity N=3,533	High familiarity N=589	Signif. diff. Source of diff.
7. Conserve and protect watersheds	3.91 *1.17*	3.97 *1.37*	3.95 *1.06*	3.67 *1.03*	

continued on next page

USDA Forest Service RMRS GTR-95. 2002

Table 43. *Continued.*

8. Preserve natural resources through policies such as no timber, no mining	3.65	3.81	3.74	3.56	
	1.31	*1.33*	*1.21*	*1.13*	
9. Protecting ecosystems and wildlife habitat	3.90	4.13	4.02	3.52	***
	1.16	*1.30*	*1.01*	*1.14*	H-L, M
10. Preserve wilderness experience	3.88	3.80	3.93	3.72	
	1.10	*1.25*	*1.06*	*0.85*	
11. Preserve local cultural uses	3.39	3.59	3.46	3.32	
	1.37	*1.25*	*1.30*	*0.98*	

[a] Standard deviation
[b] For example, L-M indicates that the source of the statistically significant difference lies in the differences between the responses of those with low agency familiarity and those with moderate familiarity
*, **, *** Mean difference is significant at   = 0.05, 0.01, 0.001

Economic Development

Survey items dealing with economic development appear in table 44.

Respondents with high or moderate agency familiarity are more in favor of restricting the development of minerals, while those with lower familiarity are more in favor of easing the permitting process. This group also feels that this is a more important role for the USDA Forest Service than do those with a higher level of agency familiarity.

More familiar respondents feel that the expansion of commercial recreation is a slightly unimportant role for the agency, while those less familiar see this as slightly important. Agency performance is rated higher (where there are differences) by those who are less familiar.

Table 44. Mean scores by respondent familiarity for economic development. Survey items are from the VOBA module of the National Survey on Recreation and the Environment.

	VOBA Objectives for the management of forests and grasslands *(1= not at all important, 5= very important)*				
Survey item number and statement	Full sample N=5,064	Low familiarity N=942	Moderate familiarity N=3,533	High familiarity N=589	Signif. diff. Source of diff.
12. Provide natural resources to dependent communities	3.60	3.67	3.58	3.31	
	1.39[a]	*1.39*	*1.39*	*1.24*	
13. Restrict development of minerals	3.96	3.51	4.14	3.97	***
	1.42	*1.87*	*1.19*	*1.08*	L-M, H[b]
14. Restrict timber harvest and grazing	3.99	4.00	4.05	3.76	
	1.27	*1.41*	*1.18*	*1.22*	
15. Make the permitting process easier for established uses	2.89	3.40	2.81	2.57	***
	1.55	*1.69*	*1.44*	*1.26*	L-M, H
16. Develop a national policy for natural resource development	4.26	4.28	4.27	4.33	
	1.23	*1.31*	*1.12*	*1.05*	
17. Expand commercial recreation	3.04	3.32	2.85	2.76	***
	1.45	*1.76*	*1.25*	*1.12*	L-M, H

continued on next page

Table 44. *Continued.*

Survey item number and statement	**VOBA Beliefs** about the role of the USDA Forest Service *(1= strongly disagree, 5= strongly agree)*				
	Full sample N=5,064	Low familiarity N=942	Moderate familiarity N=3,533	High familiarity N=589	Signif. diff. Source of diff.
12. Provide natural resources to dependent communities	3.29 *1.39*	3.46 *1.43*	3.38 *1.31*	3.12 *1.09*	
13. Restrict development of minerals	3.95 *1.43*	3.57 *1.81*	4.01 *1.34*	3.96 *1.22*	** L-M
14. Restrict timber harvest and grazing	3.94 *1.34*	3.90 *1.41*	4.08 *1.21*	3.77 *1.26*	
15. Make the permitting process easier for established uses	2.90 *1.59*	2.70 *1.81*	2.77 *1.43*	2.45 *1.40*	
16. Develop a national policy for natural resource development	4.21 *1.19*	4.04 *1.48*	4.28 *1.11*	4.05 *1.00*	* L-M
17. Expand commercial recreation	3.25 *1.53*	3.13 *1.59*	3.24 *1.44*	2.78 *1.26*	* M-H

Survey item number and statement	**VOBA Attitudes** about the performance of the Forest Service *(1= very unfavorable, 5= very favorable)*				
	Full sample N=5,064	Low familiarity N=942	Moderate familiarity N=3,533	High familiarity N=589	Signif. diff. Source of diff.
12. Provide natural resources to dependent communities	3.45 *1.30*	3.21 *1.53*	3.54 *1.16*	2.90 *1.09*	*** M-L, H
13. Restrict development of minerals	3.30 *1.50*	3.25 *1.58*	3.33 *1.49*	3.08 *1.28*	
14. Restrict timber harvest and grazing	3.50 *1.36*	3.63 *1.50*	3.42 *1.28*	3.09 *1.15*	* L-H
15. Make the permitting process easier for established uses	3.05 *1.43*	3.08 *1.75*	3.10 *1.34*	2.60 *0.97*	
16. Develop a national policy for natural resource development	3.51 *1.26*	3.71 *1.30*	3.53 *1.17*	3.46 *1.12*	
17. Expand commercial recreation	3.45 *1.24*	3.47 *1.37*	3.54 *1.21*	3.36 *0.99*	

[a] Standard deviation

[b] For example, L-M indicates that the source of the statistically significant difference lies in the differences between the responses of those with low agency familiarity and those with moderate familiarity

*, **, *** Mean difference is significant at = 0.05, 0.01, 0.001

Education

Items 18 through 23 deal with education or the dispersal of information to the public. Breakdowns by familiarity for these items are in table 45.

More familiar respondents are more in favor of informing the public about recreation concerns and about potential environmental impacts than are respondents with low familiarity. Those with moderate familiarity are the most supportive of encouraging collaboration. However, those with high familiarity see this as a more important agency role than do those with low or moderate levels of familiarity.

Respondents with low or moderate familiarity see the role of informing the public on environmental impacts and on economic values as more important than do those who have a high degree of agency familiarity.

Again, agency performance receives the highest ratings from those with less familiarity with the USDA Forest Service.

Table 45. Mean scores by respondent familiarity for education. Survey items are from the VOBA module of the National Survey on Recreation and the Environment.

| | Objectives | | | | |
| | for the management of forests and grasslands *(1= not at all important, 5= very important)* | | | | |
Survey item number and statement	Full sample N=5,064	Low familiarity N=942	Moderate familiarity N=3,533	High familiarity N=589	Signif. diff. Source of diff.
18. Develop volunteer programs to improve land (tree planting, etc.)	4.60 *0.86[a]*	4.41 *1.13*	4.62 *0.80*	4.50 *0.85*	* L-M[b]
19. Develop volunteer programs to improve facilities (trails etc.)	4.18 *1.13*	4.09 *1.47*	4.19 *1.02*	4.20 *0.88*	
20. Inform the public about recreation concerns (safety, trail etiquette, etc.)	4.53 *0.93*	4.37 *1.18*	4.57 *0.84*	4.63 *0.63*	* L-M
21. Inform the public on potential environmental impacts of uses	4.39 *1.02*	4.30 *1.13*	4.48 *0.90*	4.44 *0.81*	
22. Inform the public on economic value from developing natural resources	4.02 *1.30*	4.18 *1.34*	4.03 *1.20*	3.95 *1.13*	
23. Encourage collaboration between groups to share information	4.15 *1.20*	3.97 *1.46*	4.27 *1.05*	3.81 *1.10*	*** M-L, H

| | Beliefs | | | | |
| | about the role of the USDA Forest Service *(1= strongly disagree, 5= strongly agree)* | | | | |
Survey item number and statement	Full sample N=5,064	Low familiarity N=942	Moderate familiarity N=3,533	High familiarity N=589	Signif. diff. Source of diff.
18. Develop volunteer programs to improve land (tree planting, etc.)	4.46 *1.03*	4.41 *1.30*	4.54 *0.85*	4.52 *0.83*	
19. Develop volunteer programs to improve facilities (trails etc.)	4.22 *1.09*	4.24 *1.15*	4.28 *0.99*	4.14 *0.85*	
20. Inform the public about recreation concerns (safety, trail etiquette, etc.)	4.50 *0.93*	4.57 *0.91*	4.53 *0.91*	4.47 *0.61*	
21. Inform the public on potential environmental impacts of uses	4.48 *1.00*	4.46 *1.07*	4.46 *0.94*	4.13 *1.00*	* H-L, M
22. Inform the public on economic value from developing natural resources	4.08 *1.17*	3.91 *1.37*	4.12 *1.12*	3.76 *1.06*	* M-H
23. Encourage collaboration between groups to share information	4.15 *1.17*	3.84 *1.48*	4.20 *1.03*	4.42 *0.76*	*** L-M, H

| | Attitudes | | | | |
| | about the performance of the Forest Service *(1= very unfavorable, 5= very favorable)* | | | | |
Survey item number and statement	Full sample N=5,064	Low familiarity N=942	Moderate familiarity N=3,533	High familiarity N=589	Signif. diff. Source of diff.
18. Develop volunteer programs to improve land (tree planting, etc.)	3.85 *1.25*	4.16 *1.41*	3.82 *1.17*	3.30 *1.13*	*** L-M, H; M-H
19. Develop volunteer programs to improve facilities (trails etc.)	3.79 *1.21*	4.09 *1.14*	3.82 *1.09*	3.57 *1.02*	** L-M, H
20. Inform the public about recreation concerns (safety, trail etiquette, etc.)	3.89 *1.31*	3.95 *1.48*	3.89 *1.21*	3.83 *1.05*	

continued on next page

USDA Forest Service RMRS-GTR-95. 2002   57

Table 45. *Continued.*

	Full	Low	Moderate	High	Signif.
21. Inform the public on potential environmental impacts of uses	3.50 _1.33_	3.66 _1.45_	3.49 _1.23_	3.02 _1.08_	** H-L, M
22. Inform the public on economic value from developing natural resources	3.40 _1.38_	3.51 _1.38_	3.43 _1.32_	2.63 _1.08_	*** H-L, M
23. Encourage collaboration between groups to share information	3.72 _1.22_	3.75 _1.25_	3.75 _1.15_	3.52 _1.07_	

[a] Standard deviation

[b] For example, L-M indicates that the source of the statistically significant difference lies in the differences between the responses of those with low agency familiarity and those with moderate familiarity

*, **, *** Mean difference is significant at = 0.05, 0.01, 0.001

Natural Resource Management

Natural resource management issues are captured in the final set of items presented in table 46.

People who are more familiar with the USDA Forest Service are more in favor of making management decisions at the local level. At the same time they are less supportive of collecting an entrance fee for National Forests and Grasslands and for increased law enforcement on National Forest System lands. This group also sees the role of increasing law enforcement as less important for the agency than the less familiar groups.

Respondents with low agency familiarity see increasing the total acreage of the National Forest System as a less important role for the agency than do moderately or highly familiar individuals. Higher levels of agency familiarity again correspond to lower ratings of agency performance.

Table 46. Mean scores by respondent familiarity for natural resource management. Survey items are from the VOBA module of the National Survey on Recreation and the Environment.

	Objectives for the management of forests and grasslands *(1= not at all important, 5= very important)*				
Survey item number and statement	Full sample N=5,064	Low familiarity N=942	Moderate familiarity N=3,533	High familiarity N=589	Signif. diff. Source of diff.
24. Using public advisory committees to advise on management issues	3.90 _1.20[a]_	3.94 _1.18_	3.86 _1.17_	3.77 _1.09_	
25. Allow for diverse uses	4.07 _1.18_	3.94 _1.42_	4.15 _1.05_	3.91 _0.91_	*
26. Make management decisions at the local level (rather than national)	3.93 _1.22_	4.00 _1.30_	3.87 _1.14_	4.18 _0.99_	
27. Increase total number of acres in the National Forest and Grassland system	3.81 _1.42_	4.02 _1.36_	3.66 _1.38_	3.59 _1.23_	* L-M[b]
28. Collect an entry fee to support National Forests and Grasslands	3.66 _1.36_	3.43 _1.76_	3.81 _1.21_	2.94 _1.23_	*** M-L, H
29. Increasing law enforcement on National Forests and Grasslands	4.01 _1.21_	4.16 _1.18_	3.88 _1.22_	3.77 _0.89_	* L-M
30. Trade public lands for private to eliminate inholdings, acquire unique lands	3.05 _1.48_	2.97 _1.66_	2.98 _1.50_	3.15 _1.13_	

continued on next page

Table 46. *Continued.*

	Beliefs about the role of the USDA Forest Service *(1= strongly disagree, 5= strongly agree)*				
Survey item number and statement	Full sample N=5,064	Low familiarity N=942	Moderate familiarity N=3,533	High familiarity N=589	Signif. diff. Source of diff.
24. Using public advisory committees to advise on management issues	3.84 *1.25*	3.75 *1.41*	3.83 *1.19*	3.84 *1.14*	
25. Allow for diverse uses	4.02 *1.14*	3.96 *1.18*	4.04 *1.12*	4.13 *0.90*	
26. Make management decisions at the local level (rather than national)	3.88 *1.34*	3.73 *1.63*	3.93 *1.22*	3.57 *1.21*	*
27. Increase total number of acres in the National Forest and Grassland system	3.95 *1.47*	3.57 *1.78*	4.06 *1.26*	3.69 *1.22*	*** L-M
28. Collect an entry fee to support National Forests and Grasslands	3.69 *1.43*	3.84 *1.54*	3.74 *1.26*	3.31 *1.19*	* H-L, M
29. Increasing law enforcement on National Forests and Grasslands	4.01 *1.26*	4.12 *1.30*	4.02 *1.17*	3.47 *1.29*	** H-L, M
30. Trade public lands for private to eliminate inholdings, acquire unique lands	3.22 *1.49*	3.38 *1.76*	3.16 *1.44*	3.44 *1.18*	

	Attitudes about the performance of the Forest Service *(1= very unfavorable, 5= very favorable)*				
Survey item number and statement	Full sample N=5,064	Low familiarity N=942	Moderate familiarity N=3,533	High familiarity N=589	Signif. diff. Source of diff.
24. Using public advisory committees to advise on management issues	3.36 *1.26*	3.68 *1.14*	3.38 *1.19*	3.06 *0.91*	* L-H
25. Allow for diverse uses	3.73 *1.18*	3.91 *1.10*	3.82 *1.10*	3.44 *0.98*	* H-L, M
26. Make management decisions at the local level (rather than national)	3.49 *1.32*	3.87 *1.54*	3.45 *1.25*	3.30 *1.19*	** L-M, H
27. Increase total number of acres in the National Forest and Grassland system	3.52 *1.45*	3.61 *1.54*	3.45 *1.28*	3.00 *1.12*	* H-L, M
28. Collect an entry fee to support National Forests and Grasslands	3.61 *1.36*	3.85 *1.41*	3.52 *1.30*	3.24 *0.88*	* L-M, H
29. Increasing law enforcement on National Forests and Grasslands	3.85 *1.27*	4.10 *1.31*	3.91 *1.18*	3.48 *1.08*	** H-L, M
30. Trade public lands for private to eliminate inholdings, acquire unique lands	3.22 *1.27*	3.15 *1.59*	3.29 *1.25*	3.19 *1.13*	

[a] Standard deviation

[b] For example, L-M indicates that the source of the statistically significant difference lies in the differences between the responses of those with low agency familiarity and those with moderate familiarity

*, **, *** Mean difference is significant at = 0.05, 0.01, 0.001

Selected References

Antil, J. H. 1984. Socially responsible consumers: profile and implications for public policy. Journal of Macromarketing. (Fall): 18–39.

Antil, J. H.; Bennett, P. D. 1979. Construction and validation of a scale to measure socially responsible consumer behavior. In: Henion, K. E., II; Kinnear, T. C., eds. The conserver society. American Marketing Association: 51–68.

Austin, J. T.; Vancouver, J. B. 1996. Goal constructs in psychology: structure, process, and content. Psychological Bulletin. 120(3): 338–75.

Barron, F. H.; Barrett, B. E. 1996. Decision quality using ranked attribute weights. Management Science. 42(11): 1515–1523.

Bearden, W. O.; Netemeyer, R. G.; Mobley, M. F. 1993. Handbook of marketing scales: multi-item measures for marketing and consumer behavior research. London: Sage Publications.

Beatty, S. E.; Kahle, L. R.; Homer, P.; Misra, S. 1985. Alternative measurement approaches to consumer values: the list of values and the Rokeach value survey. Psychology and Marketing. 2(3): 181–200.

Bengston, D. N. 1993. Changing forest values and ecosystem management. Society and Natural Resources. 7: 515–33.

Braithwaite, V. A.; Scott, W. A. 1991. Values, in measures of personality and social psychological attitudes. In: Robinson, J. P.; Shaver, P. R.; Wrightsman, L. S., eds. Measures of personality and social psychological attitudes. San Diego CA: Academic Press: 661–748.

Clemen, R. T. 1996. Making hard decisions: an introduction to decision analysis. Second Edition. Belmont, CA: Duxbury Press.

Churchill, G. A., Jr. 1979. A paradigm for developing better measures of marketing constructs. Journal of Marketing Research. 16(February): 64–73.

Crowne, D. P.; Marlowe, D. 1960. A new scale of social desirability independent of psychopathology. Journal of Consulting Psychology. 24(3): 349–354.

Csikszentmihalyi, M.; Rochberg-Halton, E. 1981. The meaning of things: domestic symbols and the self. Cambridge MA: Cambridge University Press.

Deaton, A.; Muellbauer, J. 1980. Economics and consumer behavior. Cambridge, MA: Cambridge University Press.

Duffield, J. W.; Brown, T. C.; Allen, S. D. 1994. Economic value of instream flow in Montana s Big Hole and Bitterroot Rivers. Gen. Tech. Rep. RM-317. Fort Collins, CO: U.S. Department of Agriculture, Forest Service, Rocky Mountain Forest and Range Experiment Station.

Dunlap, R. E.; Van Liere, S. D. 1978. The new environmental paradigm. Journal of Environmental Education. 9(4): 10–19.

Dunlap, R. E.; Heffernan, R. B. 1975. Outdoor recreation and environmental concern: an empirical examination. Rural Sociology. 40: 18–30.

Duvall, W.; Sessions, G. 1986. Deep ecology. Salt Lake City, UT: Peregrine Smith Books.

Eagley, A. H.; Chaiken, S. 1993. The psychology of attitudes. Orlando, FL: Harcourt Brace Jovanovich College Publishers.

Edwards, W. E. 1977. How to use multiattribute utility measurement for social decisionmaking. IEEE Transactions on Systems, Man, and Cybernetics. 7: 326–340.

Fazio, R. 1986. How do attitudes guide behaviors? In: Sorrentino, R. M.; Higgins, E. T., eds. The handbook of motivation and cognition: foundations of social behavior. New York: Guilford Press: 204–43.

Fishburn, P. C. 1967. Methods of estimating additive utilities. Management Science. 13: 435–453.

Freeman, A. M. 1993. The measurement of environmental and resource values: theory and methods. Washington, DC: Resources for the Future.

Gorsuch, R. L. 1983. Factor analysis. Second Edition. Hillsdale, NJ: LEA.

Gottleib, A. 1989. The wise use agenda: a citizen's policy guide to environmental resource issues. Bellevue, WA: Free Enterprise Press.

Guttman, J. 1982. A means-end chain model based on consumer categorization processes. Journal of Marketing. 46(2): 60–72.

Gregory, R. L.; Keeney, R. L. 1994. Creating policy alternatives using stakeholder values. Management Science. 40: 1035.

Homer, P. M.; Kahle, L. R. 1988. A structural equation test of the value-attitude-behavior hierarchy. Journal of Personality and Social Psychology. 54(4): 638–46.

Hyberg, B.T. 1987. Multiattribute decision theory and forest management: a discussion and application. Forest Science. 33: 835–845.

Jones, E. S.; Calloway, W. 1995. Neutral bystander, intrusive micromanager, or useful catalyst: the role of Congress in effecting change within the Forest Service. Policy Studies Journal. 23: 4.

Kangas, J. 1994. An approach to public participation in strategic forest management planning. Forest Ecology and Management. 70: 75–88.

Keeney, R. L. 1992. Value-focused thinking: a path to creative decisionmaking. Cambridge, MA: Harvard University Press.

Keeney, R. L.; Raiffa, H. 1976. Decisions with multiple objectives. New York: John Wiley & Sons. [Republished in 1993.]

Lichtenstein, E.; Brewer, W. F. 1983. Memory for goal-directed events. Cognitive Psychology. 12: 412–45.

Manzer, L. L.; Miller, S. J. 1982. An examination of value-attitude structure in the study of donor behavior. Advances in Consumer Research. 10: 204–6.

Martin, I. M.; Bender, H. W.; Martin, W. A.; Shields, D. J. 1998. The impact of goals on the "values à attitudes à behaviors" framework. In: Proceedings: Decision Science Institute; 1998 November; Las Vegas, NV: 196–229.

Martin, W. E.; Bender, H. W.; Shields, D. J. 2000. Stakeholder objectives for public lands: rankings of forest management alternatives. Journal of Environmental Management. 58(1): 21–32.

Martin, W. E.; Bender, H. W. 1999. Modeling public land use decisions as a cooperative game. International Journal of Environment & Pollution. 12(2/3): 217–231.

McCracken, G. 1986. Culture and consumption: a theoretical account of the structure and movement of cultural meaning of consumer goods. Journal of Consumer Research. 13(June): 71–84.

Munson, J. M. 1984. Personal values: considerations on their measurement and application to five areas of research inquiry. In: Pitts, R. E.; Woodside, A. G., eds. Personal values and consumer psychology. Lexington, MA: Lexington Books: 13–33.

Nunally, J. C. 1978. Psychometric theory. New York: McGraw-Hill.

Petulla, J. M. 1980. American environmentalism: values, tactics, priorities. College Station: Texas A & M University Press.

Pfanzagl, J. 1968. Theory of measurement. New York: John Wiley & Sons.

Pykalainen, J.; Loikkanen, T. 1997. An application of numeric decision analysis on participatory forest planning: the case of Kainuu. EFI Proceedings. 14:125–132.

Ray, P. H. 1997. The emerging culture. American Demographics. (Feb): 29–34.

Reynolds, T. J.; Guttman, J. 1988. Laddering theory, method, analysis, and interpretation. Special Issue: Values in Journal of Advertising Research. 28(March): 11–31.

Richins, M. L. 1994. Valuing things: the public and private meanings of possessions. Journal of Consumer Research. 21(Dec.): 504–21.

Richins, M. L.; Dawson, S. 1992. A consumer values orientation for materialism and its measurement: scale development and validation. Journal of Consumer Research. 21(Dec.): 504–21.

Rokeach, M. J. 1968. The role of values in public opinion research. Public Opinion Quarterly. 32(Winter): 547–49.

Rokeach, M. J. 1973. The nature of human values. New York: The Free Press.

Sagoff, M. 1988. The economy of the earth: philosophy, law and the environment. Cambridge, MA: Cambridge University Press.

Sedjo, R. A. 1995. Ecosystem management: an uncharted path for public forests. Resources. 129(Fall).

Steel, B. S.; List, P.; Shindler, B. 1994. Conflicting values about federal forests: a comparison of national and Oregon publics. Society and Natural Resources. 7: 137–53.

Torgerson, W. S. 1958. Theory and methods of scaling. New York: John Wiley & Sons.

Von Winterfeldt, D.; Edwards, W. 1986. Decision analysis and behavior research. New York: Cambridge University Press.

Wilkinson, C. F.; Anderson, H. M. 1987. Land and resource planning in the National Forests. Washington, DC: Island Press.

Appendix A. Methodology

Survey Design and Implementation

The individual values, objectives, beliefs, and attitudes statements in the survey are sets of scale items that have been subjected to extensive pretesting and have been applied in various other studies. The Public Lands Values (Values) scale was developed using approximately 200 items that, through a series of iterations using both student samples and adult samples around the United States, was reduced to a 25-item scale. This scale was designed to focus on values that people hold for the environment in general and public lands in particular. It has been tested on four National Forests in Colorado (Arapaho, Roosevelt, Pike, and San Isabel) using various traditional and nontraditional stakeholder groups. For an in-depth explanation of how the Values scale was constructed and validated see Martin, Martin, Shields, and Wise (1999).

The Objectives, Beliefs, and Attitudes scale items were developed using a similar approach on a nationwide basis. First, a series of 80 focus groups were held with both traditional and nontraditional stakeholders around the continental United States. An objectives hierarchy process was conducted with each focus group with the purpose of eliciting information on people's goals for the management of forests and grasslands. The hierarchies were validated with each respective group, then merged and duplicate objectives eliminated. The fundamental objectives from the merged hierarchy were the source of almost all the objectives statements in the survey instrument. In addition, objectives that related to a few issues of importance to the USDA Forest Service, but not raised by any focus group, were also included. The final Objectives scale (Objectives for Managing Public and Private Forest and Grasslands) contains 30 items and is very similar to the set of objectives that were developed through focus groups on the Arapaho and Roosevelt, and the Pike and San Isabel, National Forests. (For any questions concerning the methodology used to develop any of the scale items please see Martin, et al. 1998; Martin, Bender, & Shields 2000).

For each objectives scale item there is a corresponding belief concerning the role of the USDA Forest Service in achieving that objective and an attitude regarding the job that the USDA Forest Service is doing in achieving that objective. The purpose of the Beliefs scale items is to ascertain whether a respondent thinks it is an appropriate role of the USDA Forest Service to manage so as to achieve a given objective on public lands. The purpose of the Attitudes scale items is to determine how good a job the respondent thinks the USDA Forest Service is doing in fulfilling the related objective, irrespective of the beliefs about whether the agency should in fact be attempting to fulfill that objective. Thus the Beliefs and Attitudes scale items tier down from the Objectives scale items, which in turn are derived from information provided by the focus groups. All scale items were measured using Likert scales of 1 to 5. The objectives scale items were anchored by 1=not at all important and 5=very important, and the values and beliefs, and scale items were anchored by 1=strongly disagree and 5=strongly agree. The attitudes scale items were anchored by 1=very unfavorable ad 5=very favorable. For all the scale items, 8= don't know and 9=refused.

This survey of the American public was administered for the USDA Forest Service via telephone by the University of Tennessee. Due to restrictions imposed by the Office of Management and Budget, contact time was limited to an average of 15 minutes. The VOBA portion of the survey was limited to 7 minutes on average, with the remainder of the time used to ask respondents the familiarity, demographic, and selected recreation-oriented questions. Because the full VOBA survey could not be completed within the allotted time window, a split sampling design was utilized. Each respondent was asked a subset of the full set of questions. Approximately 7,000 responses will be needed to ensure that the number of responses to each individual scale item are adequate to support multivariate statistical analysis. The first 5,064 NSRE responses were collected in the fall of 1999; the remaining 2,000 responses were collected during the spring and summer of 2000.

Data Analysis

This report is based on 7,069 responses. The sample data do not reflect the demographic makeup of the United States. Therefore the data were weighted so as to be consistent with the demographics of the United States for 1999 as predicted by the US Census Bureau. An overall weight was constructed for each individual response based on the following demographic factors: age, race, sex, education level, and metropolitan or non-metropolitan area. Descriptive statistics were then calculated for each of the 115 items, including mean, standard deviation, and frequency distribution. In essence, those responses from under-sampled areas or population groups were given greater than proportional weight during the calculation of the descriptive statistics for each VOBA item, while responses from over-sampled groups or areas were given less than proportional weight.

Based on past research and testing, responses to the Public Lands Values scale load on two latent variables, or factors, that have been labeled Socially Responsible Individual Values (SRIV) and Socially Responsible Management Values (SRMV). These factors were confirmed using a subset of NSRE responses where the individuals were asked the complete set of values items. Factor scores (group means) have been calculated for the SRIV and SRMV and, where appropriate, items have been reverse scored.

When the VOBA was designed, care was taken to avoid the appearance of an instrument that was biased toward or against a specific position. To do this, the "direction" of the scale varied. For example, for one item a "strongly agree" response might indicate a conservation/ preservation orientation, while for another item the same response might indicate a development orientation. While this is useful to increase the acceptance of the instrument and subsequent response rates, it creates problems when items with the opposite direction are grouped.

In order to calculate a group mean for two or more items that have the opposite direction, it is necessary to make the items move in the same direction. To illustrate this we will use an example. Suppose we want to examine the overall preference for sweets as indicated by the preference for ice cream and pie. We have two scale items. For each, 1 indicates "strongly disagree" and 5 indicates "strongly agree" as in the Public Lands Values scale. In order to avoid the appearance of bias toward or against sweets, the two items move in opposite directions: "I like ice cream" and "I don't like pie." Clearly a person who likes all sweets will answer 5 to the first item and 1 to the second. Conversely, someone who does not like sweets will answer 1 to the first and 5 to the second. If these items are grouped, these two (clearly different) respondents will have the same mean for the item group (3). This would give a researcher little (if any) useful information. In order to calculate a useful mean, we choose one of the items, in this example we'll choose the second, and reverse the scoring. So, an answer of 5 to "I don't like pie" becomes a 1 (and we can reword the item as "I like pie"). An answer of 4 becomes 2, 3 remains the same (neutral), 2 becomes 4 and 1 becomes 5. This in effect creates a new item that corresponds in direction to "I like ice cream." Respondent one (the sweet tooth) will have a mean for the two items of 5 and then respondent one's mean becomes 1. Now we have an indication of each respondent's preference for sweets. We can also calculate the mean for the entire sample and gain information about the sample's overall preference for sweets. A similar re-scoring was done for certain items in the VOBA in order to more accurately characterize overall preferences for item groups.

The objectives, beliefs, attitudes, and values items are also grouped into sections consistent with the USDA Forest Service 2000 Strategic Plan Goals and Objectives. During the development of the national survey, the focus groups were not asked to respond specifically to Forest Service Strategic Plan Objectives, but rather to express their own objectives. To a large degree, these two sets of objectives overlap. Where possible, objectives statements expressed by the focus groups (and used as the national survey) have been linked to the USDA Forest Service 2000 Strategic Plan Objectives in order to understand how well the public supports these agency objectives.

In addition, based upon the results of the focus group interviews, an objectives hierarchy was constructed for each group. These hierarchies indicated what each group or individual was attempting to achieve and how they would achieve each goal or objective. These objectives ranged from the very abstract strategic level to the more focused or concrete means level. The means level objectives are at the bottom of the hierarchy while the strategic objective is at the top. Fundamental objectives between the means level and the strategic level completed the hierarchies.

Each of the objectives hierarchies was confirmed with its respective group so as to ensure that it accurately reflected their goals and objectives. A combined objectives hierarchy was then constructed that included all the objectives stated by each group or individual interviewed. The result was a hierarchy that covered 5 strategic level objectives related to access, preservation/conservation, commodity development, education and natural resource management. These 5 strategic level objectives were supported by 30 higher-level fundamental objectives. In order to facilitate analysis, the results for the 30 fundamental level objectives (the survey items) have been grouped according to the strategic level objectives.

The 30 fundamental level objectives were used to develop 30 objectives statements that were used in the NSRE survey. The 30 objectives statements were divided into 5 groups based upon the strategic level objectives that the focus groups had identified. During the telephone interviews each respondent was asked one statement from each of the 5 strategic level groups in order to obtain a statistically valid sample for each statement and for each strategic level group.

Responses to the demographic questions were used to break the full data set down, first by region (East and West, divided at approximately the 100th meridian), then by metropolitan and non-metropolitan counties within each region. The weighed means for each item broken down into the subsets described above, along with national means for each category, are presented in Appendices F, G, and H.

Appendix B. Values Statements for NSRE Telephone Survey

1. People should be more concerned about how our public lands are used.
   Strongly
   disagree   1   2   3   4   5   Strongly agree

2. Natural resources must be preserved even if people must do without some products.
   Strongly
   disagree   1   2   3   4   5   Strongly agree

3. Consumers should be interested in the environmental consequences of the products they purchase.
   Strongly
   disagree   1   2   3   4   5   Strongly agree

4. I would be willing to sign a petition for an environmental cause.
   Strongly
   disagree   1   2   3   4   5   Strongly agree

5. The whole pollution issue has never upset me too much since I feel it's somewhat over-rated.
   Strongly
   disagree   1   2   3   4   5   Strongly agree

6. I have often thought that if we could just get by with a little less there would be more left for future generations.
   Strongly
   disagree   1   2   3   4   5   Strongly agree

7. Manufacturers should be encouraged to use recycled materials in their manufacturing and processing operations.
   Strongly
   disagree   1   2   3   4   5   Strongly agree

8. Future generations should be as important as the current one in the decisions about public lands.
   Strongly
   disagree   1   2   3   4   5   Strongly agree

9. I would be willing to pay five dollars more each time I use public lands for recreational purposes (for example, hiking, camping, hunting).
   Strongly
   disagree   1   2   3   4   5   Strongly agree

10. People should urge their friends to limit their use of products made from scarce resources.
    Strongly
    disagree   1   2   3   4   5   Strongly agree

11. I am glad there are national forests even if I never get to see them.
    Strongly
    disagree   1   2   3   4   5   Strongly agree

12. People can think public lands are valuable even if they do not actually go there themselves.

    Strongly
    disagree   1   2   3   4   5
    Strongly
    agree

13. I am willing to stop buying products from companies that pollute the environment even though it might be inconvenient.

    Strongly
    disagree   1   2   3   4   5
    Strongly
    agree

14. I am willing to make personal sacrifices for the sake of slowing down pollution.

    Strongly
    disagree   1   2   3   4   5
    Strongly
    agree

15. Forests have a right to exist for their own sake, regardless of human concerns and uses.

    Strongly
    disagree   1   2   3   4   5
    Strongly
    agree

16. Wildlife, plants and humans have equal rights to live and grow.

    Strongly
    disagree   1   2   3   4   5
    Strongly
    agree

17. Donating time or money to worthy causes is important to me.

    Strongly
    disagree   1   2   3   4   5
    Strongly
    agree

18. We should actively harvest more trees to meet the needs of a much larger human population.

    Strongly
    disagree   1   2   3   4   5
    Strongly
    agree

19. The most important role for the public lands is providing jobs and income for local people.

    Strongly
    disagree   1   2   3   4   5
    Strongly
    agree

20. The decision to develop resources should be based mostly on economic grounds.

    Strongly
    disagree   1   2   3   4   5
    Strongly
    agree

21. The main reason for maintaining resources today is so we can develop them in the future if we need to.

    Strongly
    disagree   1   2   3   4   5
    Strongly
    agree

22. I think that the public land managers are doing an adequate job of protecting natural resources from being over used.

    Strongly
    disagree   1   2   3   4   5
    Strongly
    agree

23. The primary use of forests should be for products that are useful to humans.

    Strongly
    disagree   1   2   3   4   5
    Strongly
    agree

24. The Federal government should subsidize the development and leasing of public lands to companies.

    Strongly
    disagree   1   2   3   4   5   Strongly
                                                     agree

25. The government has better places to spend money than devoting resources to a strong conservation program.

    Strongly
    disagree   1   2   3   4   5   Strongly
                                                     agree

Appendix C. Objectives Statements for the NSRE Phone Survey

1. Expanding access for motorized off-highway vehicles on forests and grasslands (for example, snowmobiling or 4-wheel driving).

   Not at all   Very
   important   1   2   3   4   5   important

2. Developing and maintaining continuous trail systems that cross both public and private land for motorized vehicles such as snowmobiles or ATVs.

   Not at all   Very
   important   1   2   3   4   5   important

3. Developing and maintaining continuous trail systems that cross both public and private land for non-motorized recreation such as hiking or cross-country skiing.

   Not at all   Very
   important   1   2   3   4   5   important

4. Designating some existing recreation trails for specific use (for example, creating separate trails for snowmobiling and cross-country skiing, or for mountain biking and horseback riding).

   Not at all   Very
   important   1   2   3   4   5   important

5. Developing new paved roads on forests and grasslands for access for cars and recreational vehicles.

   Not at all   Very
   important   1   2   3   4   5   important

6. Designating more wilderness areas on public land that stops access for development and motorized uses.

   Not at all   Very
   important   1   2   3   4   5   important

7. Conserving and protecting forests and grasslands that are the source of our water resources, such as streams, lakes, and watershed areas.

   Not at all   Very
   important   1   2   3   4   5   important

8. Preserving the natural resources of forests and grasslands through such policies as no timber harvesting or no mining.

   Not at all   Very
   important   1   2   3   4   5   important

9. Protecting ecosystems and wildlife habitats.

   Not at all   Very
   important   1   2   3   4   5   important

10. Preserving the ability to have a "wilderness" experience on forests and grasslands.

   Not at all   Very
   important   1   2   3   4   5   important

11. Preserving the cultural uses of forests and grasslands by Native Americans and Native Hispanics such as firewood gathering, herb/berry/plant gathering, and ceremonial access.

| Not at all important | 1 | 2 | 3 | 4 | 5 | Very important |

12. Providing natural resources from forests and grasslands to support communities dependent on grazing, mining, or timber harvesting.

| Not at all important | 1 | 2 | 3 | 4 | 5 | Very important |

13. Restricting mineral development on forests and grasslands.

| Not at all important | 1 | 2 | 3 | 4 | 5 | Very important |

14. Restricting timber harvesting and grazing on forests and grasslands.

| Not at all important | 1 | 2 | 3 | 4 | 5 | Very important |

15. Making the permitting process easier for some established uses of forests and grasslands such as grazing, logging, mining, and commercial recreation.

| Not at all important | 1 | 2 | 3 | 4 | 5 | Very important |

16. Developing a national policy that guides natural resource development of all kinds (for example, specifies levels of extraction and regulates environmental impacts).

| Not at all important | 1 | 2 | 3 | 4 | 5 | Very important |

17. Expanding commercial recreation on forests and grasslands (for example, ski areas, guide services, or outfitters).

| Not at all important | 1 | 2 | 3 | 4 | 5 | Very important |

18. Developing volunteer programs to improve forests and grasslands (for example, planting trees or improving water quality).

| Not at all important | 1 | 2 | 3 | 4 | 5 | Very important |

19. Developing volunteer programs to maintain trails and facilities on forests and grasslands (for example, trail maintenance or campground maintenance).

| Not at all important | 1 | 2 | 3 | 4 | 5 | Very important |

20. Informing the public about recreation concerns on forests and grasslands such as safety, trail etiquette, and respect for wildlife.

| Not at all important | 1 | 2 | 3 | 4 | 5 | Very important |

21. Informing the public on the potential environmental impacts of all uses associated with forests and grasslands.

| Not at all important | 1 | 2 | 3 | 4 | 5 | Very important |

22. Informing the public on the economic value received by developing our natural resources.

| Not at all important | 1 | 2 | 3 | 4 | 5 | Very important |

23. Encouraging collaboration between groups in order to share information concerning uses of forests and grasslands.

        Not at all   Very
        important   1   2   3   4   5   important

24. Using public advisory committees to advise on public land management issues.

        Not at all   Very
        important   1   2   3   4   5   important

25. Allowing for diverse uses of forests and grasslands such as grazing, recreation, and wildlife habitat.

        Not at all   Very
        important   1   2   3   4   5   important

26. Making management decisions concerning the use of forests and grasslands at the local level rather than at the national level.

        Not at all   Very
        important   1   2   3   4   5   important

27. Increasing the total number of acres in the public land system.

        Not at all   Very
        important   1   2   3   4   5   important

28. Paying an entry fee that goes to support public land.

        Not at all   Very
        important   1   2   3   4   5   important

29. Increasing law enforcement efforts by public land agencies on public lands.

        Not at all   Very
        important   1   2   3   4   5   important

30. Allowing public land managers to trade public lands for private lands (for example, to eliminate private property within public land boundaries or to acquire unique areas of land).

        Not at all   Very
        important   1   2   3   4   5   important

Appendix D. Beliefs Statements for the NSRE Phone Survey

1. Expanding access for motorized off-highway vehicles on forests and grasslands (for example, snowmobiling or 4-wheel driving).

   Strongly
   disagree   1   2   3   4   5   Strongly agree

2. Developing and maintaining continuous trail systems that cross both public and private land for motorized vehicles such as snowmobiles or ATVs.

   Strongly
   disagree   1   2   3   4   5   Strongly agree

3. Developing and maintaining continuous trail systems that cross both public and private land for non-motorized recreation such as hiking or cross-country skiing.

   Strongly
   disagree   1   2   3   4   5   Strongly agree

4. Designating some existing recreation trails for specific use (for example, creating separate trails for snowmobiling and cross-country skiing, or for mountain biking and horseback riding).

   Strongly
   disagree   1   2   3   4   5   Strongly agree

5. Developing new paved roads on forests and grasslands for access for cars and recreational vehicles.

   Strongly
   disagree   1   2   3   4   5   Strongly agree

6. Designating more wilderness areas on public land that stops access for development and motorized uses.

   Strongly
   disagree   1   2   3   4   5   Strongly agree

7. Conserving and protecting forests and grasslands that are the source of our water resources, such as streams, lakes, and watershed areas.

   Strongly
   disagree   1   2   3   4   5   Strongly agree

8. Preserving the natural resources of forests and grasslands through such policies as no timber harvesting or no mining.

   Strongly
   disagree   1   2   3   4   5   Strongly agree

9. Protecting ecosystems and wildlife habitats.

   Strongly
   disagree   1   2   3   4   5   Strongly agree

10. Preserving the ability to have a "wilderness" experience on forests and grasslands.

    Strongly
    disagree   1   2   3   4   5   Strongly agree

11. Preserving the cultural uses of forests and grasslands by Native Americans and Native Hispanics such as firewood gathering, herb/berry/plant gathering, and ceremonial access.

Strongly disagree   1   2   3   4   5   Strongly agree

12. Providing natural resources from forests and grasslands to support communities dependent on grazing, mining, or timber harvesting.

Strongly disagree   1   2   3   4   5   Strongly agree

13. Restricting mineral development on forests and grasslands.

Strongly disagree   1   2   3   4   5   Strongly agree

14. Restricting timber harvesting and grazing on forests and grasslands.

Strongly disagree   1   2   3   4   5   Strongly agree

15. Making the permitting process easier for some established uses of forests and grasslands such as grazing, logging, mining, and commercial recreation.

Strongly disagree   1   2   3   4   5   Strongly agree

16. Developing a national policy that guides natural resource development of all kinds (for example, specifies levels of extraction and regulates environmental impacts).

Strongly disagree   1   2   3   4   5   Strongly agree

17. Expanding commercial recreation on forests and grasslands (for example, ski areas, guide services, or outfitters).

Strongly disagree   1   2   3   4   5   Strongly agree

18. Developing volunteer programs to improve forests and grasslands (for example, planting trees or improving water quality).

Strongly disagree   1   2   3   4   5   Strongly agree

19. Developing volunteer programs to maintain trails and facilities on forests and grasslands (for example, trail maintenance or campground maintenance).

Strongly disagree   1   2   3   4   5   Strongly agree

20. Informing the public about recreation concerns on forests and grasslands such as safety, trail etiquette, and respect for wildlife.

Strongly disagree   1   2   3   4   5   Strongly agree

21. Informing the public on the potential environmental impacts of all uses associated with forests and grasslands.

Strongly disagree   1   2   3   4   5   Strongly agree

22. Informing the public on the economic value received by developing our natural resources.

    Strongly   Strongly
    disagree   1   2   3   4   5   agree

23. Encouraging collaboration between groups in order to share information concerning uses of forests and grasslands.

    Strongly   Strongly
    disagree   1   2   3   4   5   agree

24. Using public advisory committees to advise on public land management issues.

    Strongly   Strongly
    disagree   1   2   3   4   5   agree

25. Allowing for diverse uses of forests and grasslands such as grazing, recreation, and wildlife habitat.

    Strongly   Strongly
    disagree   1   2   3   4   5   agree

26. Making management decisions concerning the use of forests and grasslands at the local level rather than at the national level.

    Strongly   Strongly
    disagree   1   2   3   4   5   agree

27. Increasing the total number of acres in the public land system.

    Strongly   Strongly
    disagree   1   2   3   4   5   agree

28. Paying an entry fee that goes to support public land.

    Strongly   Strongly
    disagree   1   2   3   4   5   agree

29. Increasing law enforcement efforts by public land agencies on public lands.

    Strongly   Strongly
    disagree   1   2   3   4   5   agree

30. Allowing public land managers to trade public lands for private lands (for example, to eliminate private property within public land boundaries or to acquire unique areas of land).

    Strongly   Strongly
    disagree   1   2   3   4   5   agree

Appendix E. Attitudes Statements for the NSRE Phone Survey

1. Expanding access for motorized off-highway vehicles on forests and grasslands (for example, snowmobiling or 4-wheel driving).

Very unfavorable	1	2	3	4	5	Very favorable

2. Developing and maintaining continuous trail systems that cross both public and private land for motorized vehicles such as snowmobiles or ATVs.

Very unfavorable	1	2	3	4	5	Very favorable

3. Developing and maintaining continuous trail systems that cross both public and private land for non-motorized recreation such as hiking or cross-country skiing.

Very unfavorable	1	2	3	4	5	Very favorable

4. Designating some existing recreation trails for specific use (for example, creating separate trails for snowmobiling and cross-country skiing, or for mountain biking and horseback riding).

Very unfavorable	1	2	3	4	5	Very favorable

5. Developing new paved roads on forests and grasslands for access for cars and recreational vehicles.

Very unfavorable	1	2	3	4	5	Very favorable

6. Designating more wilderness areas on public land that stops access for development and motorized uses.

Very unfavorable	1	2	3	4	5	Very favorable

7. Conserving and protecting forests and grasslands that are the source of our water resources, such as streams, lakes, and watershed areas.

Very unfavorable	1	2	3	4	5	Very favorable

8. Preserving the natural resources of forests and grasslands through such policies as no timber harvesting or no mining.

Very unfavorable	1	2	3	4	5	Very favorable

9. Protecting ecosystems and wildlife habitats.

Very unfavorable	1	2	3	4	5	Very favorable

10. Preserving the ability to have a "wilderness" experience on forests and grasslands.

Very unfavorable	1	2	3	4	5	Very favorable

11. Preserving the cultural uses of forests and grasslands by Native Americans and Native Hispanics such as firewood gathering, herb/berry/plant gathering, and ceremonial access.

| Very unfavorable | 1 | 2 | 3 | 4 | 5 | Very favorable |

12. Providing natural resources from forests and grasslands to support communities dependent on grazing, mining, or timber harvesting.

| Very unfavorable | 1 | 2 | 3 | 4 | 5 | Very favorable |

13. Restricting mineral development on forests and grasslands.

| Very unfavorable | 1 | 2 | 3 | 4 | 5 | Very favorable |

14. Restricting timber harvesting and grazing on forests and grasslands.

| Very unfavorable | 1 | 2 | 3 | 4 | 5 | Very favorable |

15. Making the permitting process easier for some established uses of forests and grasslands such as grazing, logging, mining, and commercial recreation.

| Very unfavorable | 1 | 2 | 3 | 4 | 5 | Very favorable |

16. Developing a national policy that guides natural resource development of all kinds (for example, specifies levels of extraction and regulates environmental impacts).

| Very unfavorable | 1 | 2 | 3 | 4 | 5 | Very favorable |

17. Expanding commercial recreation on forests and grasslands (for example, ski areas, guide services, or outfitters).

| Very unfavorable | 1 | 2 | 3 | 4 | 5 | Very favorable |

18. Developing volunteer programs to improve forests and grasslands (for example, planting trees or improving water quality).

| Very unfavorable | 1 | 2 | 3 | 4 | 5 | Very favorable |

19. Developing volunteer programs to maintain trails and facilities on forests and grasslands (for example, trail maintenance or campground maintenance).

| Very unfavorable | 1 | 2 | 3 | 4 | 5 | Very favorable |

20. Informing the public about recreation concerns on forests and grasslands such as safety, trail etiquette, and respect for wildlife.

| Very unfavorable | 1 | 2 | 3 | 4 | 5 | Very favorable |

21. Informing the public on the potential environmental impacts of all uses associated with forests and grasslands.

| Very unfavorable | 1 | 2 | 3 | 4 | 5 | Very favorable |

22. Informing the public on the economic value received by developing our natural resources.

| Very unfavorable | 1 | 2 | 3 | 4 | 5 | Very favorable |

23. Encouraging collaboration between groups in order to share information concerning uses of forests and grasslands.

| Very unfavorable | 1 | 2 | 3 | 4 | 5 | Very favorable |

24. Using public advisory committees to advise on public land management issues.

| Very unfavorable | 1 | 2 | 3 | 4 | 5 | Very favorable |

25. Allowing for diverse uses of forests and grasslands such as grazing, recreation, and wildlife habitat.

| Very unfavorable | 1 | 2 | 3 | 4 | 5 | Very favorable |

26. Making management decisions concerning the use of forests and grasslands at the local level rather than at the national level.

| Very unfavorable | 1 | 2 | 3 | 4 | 5 | Very favorable |

27. Increasing the total number of acres in the public land system.

| Very unfavorable | 1 | 2 | 3 | 4 | 5 | Very favorable |

28. Paying an entry fee that goes to support public land.

| Very unfavorable | 1 | 2 | 3 | 4 | 5 | Very favorable |

29. Increasing law enforcement efforts by public land agencies on public lands.

| Very unfavorable | 1 | 2 | 3 | 4 | 5 | Very favorable |

30. Allowing public land managers to trade public lands for private lands (for example, to eliminate private property within public land boundaries or to acquire unique areas of land).

| Very unfavorable | 1 | 2 | 3 | 4 | 5 | Very favorable |

Appendix F. USDA Forest Service Familiarity Questions

Introduction: "We are interested in how familiar you are with the responsibilities of the United States Forest Service. Based on your knowledge of the Forest Service, please tell me if you think each of the following statements is TRUE or FALSE, or if you don't know."

FS1—The Forest Service regulates hunting and fishing seasons.

FS2—The Forest Service has Smokey Bear as its mascot.

FS3—The Forest Service enforces the Endangered Species Act.

FS4—The Forest Service manages national forests for recreation, timber, and water.

FS5—The Forest Service provides visitor information and protects wildlife in National Parks.

Appendix G. Weighted Means by Demographic Breakdowns Grouped According to USDA Forest Service Strategic Plan Objectives

Goal 1: Ecosystem Health: Promote ecosystem health and conservation using a collaborative approach to sustain the Nation's forests, grasslands, and watersheds.

Strategic Plan Objective 1.a. Improve and protect watershed conditions to provide the water quality and quantity and the soil productivity necessary to support ecological functions and intended beneficial water uses (see table G1).

Table G1. Strategic Plan Objective 1.a.: Improve and protect watershed conditions to provide the water quality and quantity and the soil productivity necessary to support ecological functions and intended beneficial water uses. Survey items are from the VOBA module of the National Survey on Recreation and the Environment.

| | | **Objectives** for the management of forests and grasslands (1= not at all important, 5= very important) | | | | | | | |
| | **Full sample** | **East** | | **West** | | **Non-metro** | | **Metro** | |
Survey item number and statement		Non-metro	Metro	Non-metro	Metro	East	West	East	West
7. Conserve and protect watersheds	4.73 0.76[a]	4.68 0.58	4.75 0.81	4.57 0.72	4.75[*] 0.86	4.68 0.58	4.57 0.72	4.75 0.81	4.75 0.86
18. Develop volunteer programs to improve land (tree planting, etc.)	4.60 0.86	4.47 0.72	4.63[**] 0.88	4.42 0.77	4.65[*] 0.98	4.47 0.72	4.42 0.77	4.63 0.88	4.65 0.98

| | | **Beliefs** about the role of the USDA Forest Service (1= strongly disagree, 5= strongly agree) | | | | | | | |
| | **Full sample** | **East** | | **West** | | **Non-metro** | | **Metro** | |
Survey item number and statement		Non-metro	Metro	Non-metro	Metro	East	West	East	West
7. Conserve and protect watersheds	4.61 0.83	4.57 0.69	4.67 0.84	4.36 0.72	4.57[*] 0.99	4.57 0.69	4.36[*] 0.72	4.67 0.84	4.57 0.99
18. Develop volunteer programs to improve land (tree planting, etc.)	4.46 1.03	4.48 0.77	4.50 1.02	4.20 0.84	4.42 1.38	4.48 0.77	4.20[*] 0.84	4.50 1.02	4.42 1.38

| | | **Attitudes** about the performance of the Forest Service (1= very unfavorable, 5= very favorable) | | | | | | | |
| | **Full sample** | **East** | | **West** | | **Non-metro** | | **Metro** | |
Survey item number and statement		Non-metro	Metro	Non-metro	Metro	East	West	East	West
7. Conserve and protect watersheds	3.91 1.17	3.81 0.93	3.91 1.23	3.71 0.91	4.00[**] 1.43	3.81 0.93	3.71 0.91	3.91 1.23	4.00 1.43
18. Develop volunteer programs to improve land (tree planting, etc.)	3.85 1.25	3.76 0.97	3.91 1.26	3.78 0.91	3.81 1.67	3.76 0.97	3.78 0.91	3.91 1.26	3.81 1.67

[a] Standard deviation
[*], [**], [***] Mean difference is significant at = 0.05, 0.01, 0.001

Strategic Plan Objective 1.b. Provide ecological conditions to sustain viable populations of native and desired nonnative species and to achieve objectives for Management Indicator Species (MIS)/focal species (see table G2).

Table G2. Strategic Plan Objective 1.b.: Provide ecological conditions to sustain viable populations of native and desired nonnative species and to achieve objectives for Management Indicator Species (MIS)/focal species. Survey items are from the VOBA module of the National Survey on Recreation and the Environment.

Survey item number and statement	Full sample	Objectives for the management of forests and grasslands (1= not at all important, 5= very important)							
		East		West		Non-metro		Metro	
		Non-metro	Metro	Non-metro	Metro	East	West	East	West
6. Designate more wilderness to stop development & motorized access	3.84 *1.41[a]*	3.70 *1.05*	3.90* *1.51*	3.28 *1.14*	3.89*** *1.66*	3.70 *1.05*	3.28** *1.14*	3.90 *1.51*	3.89 *1.66*
8. Preserve natural resources through policies such as no timber, no mining	4.22 *1.23*	3.99 *1.04*	4.29*** *1.26*	3.83 *1.05*	4.29*** *1.46*	3.99 *1.04*	3.83 *1.05*	4.29 *1.26*	4.29 *1.46*
9. Protecting ecosystems and wildlife habitat	4.58 *0.92*	4.49 *0.76*	4.61* *0.98*	4.23 *0.82*	4.62*** *1.02*	4.49 *0.76*	4.23** *0.82*	4.61 *0.98*	4.62 *1.02*
13. Restrict development of minerals	3.96 *1.42*	3.88 *1.07*	4.10* *1.40*	3.76 *1.05*	3.82 *1.95*	3.88 *1.07*	3.76 *1.05*	4.10 *1.40*	3.82** *1.95*
14. Restrict timber harvest and grazing	3.99 *1.27*	3.92 *0.98*	4.09* *1.31*	3.56 *1.05*	3.95** *1.58*	3.92 *0.98*	3.56** *1.05*	4.09 *1.31*	3.95 *1.58*

Survey item number and statement	Full sample	Beliefs about the role of the USDA Forest Service (1= strongly disagree, 5= strongly agree)							
		East		West		Non-metro		Metro	
		Non-metro	Metro	Non-metro	Metro	East	West	East	West
6. Designate more wilderness to stop development & motorized access	3.66 *1.46*	3.58 *1.06*	3.84* *1.54*	3.09 *1.22*	3.54* *1.75*	3.58 *1.06*	3.09** *1.22*	3.84 *1.54*	3.54* *1.75*
8. Preserve natural resources through policies such as no timber, no mining	4.21 *1.27*	4.05 *1.01*	4.29* *1.30*	3.83 *1.06*	4.22* *1.57*	4.05 *1.01*	3.83 *1.06*	4.29 *1.30*	4.22 *1.57*
9. Protecting ecosystems and wildlife habitat	4.53 *0.98*	4.41 *0.78*	4.55 *1.08*	4.40 *0.69*	4.56 *1.14*	4.41 *0.78*	4.40 *0.69*	4.55 *1.08*	4.56 *1.14*
13. Restrict development of minerals	3.95 *1.43*	3.80 *1.11*	4.06* *1.48*	3.55 *0.95*	3.91 *1.80*	3.80 *1.11*	3.55 *0.95*	4.06 *1.48*	3.91 *1.80*
14. Restrict timber harvest and grazing	3.94 *1.34*	3.71 *1.13*	4.02* *1.39*	3.09 *1.23*	4.07*** *1.47*	3.71 *1.13*	3.09*** *1.23*	4.02 *1.39*	4.07 *1.47*

continued on next page

Table G2. *Continued.*

		Attitudes about the performance of the Forest Service *(1= very unfavorable, 5= very favorable)*							
Survey item number and statement	**Full sample**	**East**		**West**		**Non-metro**		**Metro**	
		Non-metro	Metro	Non-metro	Metro	East	West	East	West
6. Designate more wilderness to stop development & motorized access	3.45 *1.34*	3.34 *1.02*	3.35 *1.46*	3.24 *0.95*	3.65** *1.59*	3.34 *1.02*	3.24 *0.95*	3.35 *1.46*	3.65** *1.59*
8. Preserve natural resources through policies such as no timber, no mining	3.65 *1.31*	3.61 *1.03*	3.67 *1.38*	3.32 *0.90*	3.69** *1.63*	3.61 *1.03*	3.32* *0.90*	3.67 *1.38*	3.69 *1.63*
9. Protecting ecosystems and wildlife habitat	3.90 *1.16*	4.06 *0.80*	3.89* *1.28*	3.97 *0.83*	3.82 *1.42*	4.06 *0.80*	3.97 *0.83*	3.89 *1.28*	3.82 *1.42*
13. Restrict development of minerals	3.30 *1.50*	3.57 *1.04*	3.31** *1.61*	3.13 *0.96*	3.21 *1.98*	3.57 *1.04*	3.13** *0.96*	3.31 *1.61*	3.21 *1.98*
14. Restrict timber harvest and grazing	3.50 *1.36*	3.34 *1.04*	3.55* *1.46*	2.96 *1.01*	3.61*** *1.66*	3.34 *1.04*	2.96** *1.01*	3.55 *1.46*	3.61 *1.66*

[a] Standard deviation

*, **, *** Mean difference is significant at = 0.05, 0.01, 0.001

Strategic Plan Objective 1.c. Increase the amount of forests and grasslands restored to or maintained in a healthy condition with reduced risk and damage from fires, insects and diseases, and invasive species (see table G3).

Table G3. Strategic Plan Objective 1.c.: Increase the amount of forests and grasslands restored to or maintained in a healthy condition with reduced risk and damage from fires, insects and diseases, and invasive species. Survey items are from the VOBA module of the National Survey on Recreation and the Environment.

		Objectives for the management of forests and grasslands *(1= not at all important, 5= very important)*							
Survey item number and statement	**Full samples**	**East**		**West**		**Non-metro**		**Metro**	
		Non-metro	Metro	Non-metro	Metro	East	West	East	West
9. Protecting ecosystems and wildlife habitat	4.58 *0.92[a]*	4.49 *0.76*	4.61* *0.98*	4.23 *0.82*	4.62*** *1.02*	4.49 *0.76*	4.23** *0.82*	4.61 *0.98*	4.62 *1.02*
18. Develop volunteer programs to improve land (tree planting, etc.)	4.60 *0.86*	4.47 *0.72*	4.63** *0.88*	4.42 *0.77*	4.65* *0.98*	4.47 *0.72*	4.42 *0.77*	4.63 *0.88*	4.65 *0.98*

		Beliefs about the role of the USDA Forest Service *(1= strongly disagree, 5= strongly agree)*							
Survey item number and statement	**Full samples**	**East**		**West**		**Non-metro**		**Metro**	
		Non-metro	Metro	Non-metro	Metro	East	West	East	West
9. Protecting ecosystems and wildlife habitat	4.53 *0.98*	4.41 *0.78*	4.55 *1.08*	4.40 *0.69*	4.56 *1.14*	4.41 *0.78*	4.40 *0.69*	4.55 *1.08*	4.56 *1.14*

continued on next page

Table G3. *Continued.*

	Full samples	East		West		Non-metro		Metro	
18. Develop volunteer programs to improve land (tree planting, etc.)	4.46 *1.03*	4.48 *0.77*	4.50 *1.02*	4.20 *0.84*	4.42 *1.38*	4.48 *0.77*	4.20* *0.84*	4.50 *1.02*	4.42 *1.38*

		Attitudes							
		about the performance of the Forest Service *(1= very unfavorable, 5= very favorable)*							
Survey item number and statement	Full samples	East		West		Non-metro		Metro	
		Non-metro	Metro	Non-metro	Metro	East	West	East	West
9. Protecting ecosystems and wildlife habitat	3.90 *1.16*	4.06 *0.80*	3.89* *1.28*	3.97 *0.83*	3.82 *1.42*	4.06 *0.80*	3.97 *0.83*	3.89 *1.28*	3.82 *1.42*
18. Develop volunteer programs to improve land (tree planting, etc.)	3.85 *1.25*	3.76 *0.97*	3.91 *1.26*	3.78 *0.91*	3.81 *1.67*	3.76 *0.97*	3.78 *0.91*	3.91 *1.26*	3.81 *1.67*

[a] Standard deviation
*, **, *** Mean difference is significant at = 0.05, 0.01, 0.001

Goal 2: Multiple Benefits to People: Provide a variety of uses, values, products, and services for present and future generations by managing within the capability of sustainable ecosystems.

Strategic Plan Objective 2.a. Improve the capability of the Nation's forests and grasslands to provide diverse, high-quality outdoor recreation opportunities (see table G4).

Table G4. Strategic Plan Objective 2.a.: Improve the capability of the Nation's forests and grasslands to provide diverse, high-quality outdoor recreation opportunities. Survey items are from the VOBA module of the National Survey on Recreation and the Environment.

		Objectives							
		for the management of forests and grasslands *(1= not at all important, 5= very important)*							
Survey item number and statement	Full sample	East		West		Non-metro		Metro	
		Non-metro	Metro	Non-metro	Metro	East	West	East	West
1. Expand off-highway motorized access	2.41 *1.49[a]*	2.48 *1.11*	2.28* *1.55*	2.43 *0.98*	2.57 *1.97*	2.48 *1.11*	2.43 *0.98*	2.28 *1.55*	2.57** *1.97*
2. Trails for motorized vehicles	2.82 *1.47*	3.06 *1.11*	2.79** *1.60*	2.92 *0.98*	2.76 *1.75*	3.06 *1.11*	2.92 *0.98*	2.79 *1.60*	2.76 *1.75*
3. Trails for non-motorized recreation	3.75 *1.33*	3.56 *0.92*	3.80** *1.43*	3.27 *0.89*	3.82*** *1.67*	3.56 *0.92*	3.27* *0.89*	3.80 *1.43*	3.82 *1.67*
4. Designate trails for specific use	3.59 *1.43*	3.72 *0.98*	3.65 *1.63*	3.83 *1.01*	3.36** *1.63*	3.72 *0.98*	3.83 *1.01*	3.65 *1.63*	3.36** *1.63*
5. Develop new paved roads	2.62 *1.49*	2.59 *1.14*	2.57 *1.64*	2.45 *1.05*	2.76* *1.80*	2.59 *1.14*	2.45 *1.05*	2.57 *1.64*	2.76 *1.80*
17. Expand commercial recreation	3.04 *1.45*	3.02 *1.00*	2.87 *1.49*	3.18 *1.03*	3.31 *1.95*	3.02 *1.00*	3.18 *1.03*	2.87 *1.49*	3.31*** *1.95*
19. Develop volunteer programs to improve facilities (trails, etc.)	4.18 *1.13*	4.04 *0.87*	4.09 *1.31*	4.32 *0.66*	4.34 *1.24*	4.04 *0.87*	4.32** *0.66*	4.09 *1.31*	4.34*** *1.24*

continued on next page

Table G4. *Continued.*

Survey item number and statement	Full sample	Non-metro	Metro	Non-metro	Metro	East	West	East	West
20. Inform public about recreation concerns (safety, trail etiquette, etc.)	4.53 *0.93*	4.47 *0.77*	4.58[*] *0.97*	4.35 *0.83*	4.51[*] *1.08*	4.47 *0.77*	4.35 *0.83*	4.58 *0.97*	4.51 *1.08*
28. Collect an entry fee to support National Forests and Grasslands	3.66 *1.36*	3.69 *0.99*	3.65 *1.51*	3.67 *1.22*	3.66 *1.52*	3.69 *0.99*	3.67 *1.22*	3.65 *1.51*	3.66 *1.52*

Beliefs
about the role of the USDA Forest Service
(1= strongly disagree, 5= strongly agree)

Survey item number and statement	Full sample	East Non-metro	East Metro	West Non-metro	West Metro	Non-metro East	Non-metro West	Metro East	Metro West
1. Expand off-highway motorized access	2.52 *1.50*	2.67 *1.12*	2.50 *1.61*	2.70 *1.17*	2.47 *1.76*	2.67 *1.12*	2.70 *1.17*	2.50 *1.61*	2.47 *1.76*
2. Trails for motorized vehicles	2.98 *1.51*	2.95 *1.16*	2.86 *1.71*	2.84 *1.04*	3.22 *1.72*	2.95 *1.16*	2.84 *1.04*	2.86 *1.71*	3.22[**] *1.72*
3. Trails for non-motorized recreation	3.71 *1.37*	3.45 *1.07*	3.75[*] *1.49*	3.19 *0.96*	3.82[**] *1.59*	3.45 *1.07*	3.19 *0.96*	3.75 *1.49*	3.82 *1.59*
4. Designate trails for specific use	3.94 *1.25*	4.11 *0.93*	3.92 *1.34*	4.00 *0.92*	3.88 *1.53*	4.11 *0.93*	4.00 *0.92*	3.92 *1.34*	3.88 *1.53*
5. Develop new paved roads	2.70 *1.57*	2.65 *1.05*	2.53 *1.67*	2.24 *0.95*	3.02[**] *2.05*	2.65 *1.05*	2.24[*] *0.95*	2.53 *1.67*	3.02[***] *2.05*
17. Expand commercial recreation	3.25 *1.53*	3.20 *1.05*	3.15 *1.63*	2.87 *1.00*	3.48[**] *1.97*	3.20 *1.05*	2.87[*] *1.00*	3.15 *1.63*	3.48[***] *1.97*
19. Develop volunteer programs to improve facilities (trails, etc.)	4.22 *1.09*	4.18 *0.85*	4.20 *1.18*	4.18 *0.82*	4.28 *1.31*	4.18 *0.85*	4.18 *0.82*	4.20 *1.18*	4.28 *1.31*
20. Inform public about recreation concerns (safety, trail etiquette, etc.)	4.50 *0.93*	4.50 *0.71*	4.55 *1.00*	4.42 *0.72*	4.45 *1.12*	4.50 *0.71*	4.42 *0.72*	4.55 *1.00*	4.45 *1.12*
28. Collect an entry fee to support National Forests and Grasslands	3.69 *1.43*	3.64 *0.95*	3.81 *1.49*	3.51 *0.89*	3.61 *1.93*	3.64 *0.95*	3.51 *0.89*	3.81 *1.49*	3.61 *1.93*

Attitudes
about the performance of the Forest Service
(1= very unfavorable, 5= very favorable)

Survey item number and statement	Full sample	East Non-metro	East Metro	West Non-metro	West Metro	Non-metro East	Non-metro West	Metro East	Metro West
1. Expand off-highway motorized access	2.97 *1.45*	3.06 *1.03*	2.95 *1.61*	2.97 *1.00*	2.95 *1.82*	3.06 *1.03*	2.97 *1.00*	2.95 *1.61*	2.95 *1.82*
2. Trails for motorized vehicles	3.25 *1.38*	3.24 *0.94*	3.24 *1.51*	3.24 *0.96*	3.29 *1.71*	3.24 *0.94*	3.24 *0.96*	3.24 *1.51*	3.29 *1.71*
3. Trails for non-motorized recreation	3.59 *1.27*	3.54 *0.93*	3.53 *1.36*	3.34 *1.09*	3.74[**] *1.57*	3.54 *0.93*	3.34 *1.09*	3.53 *1.36*	3.74[*] *1.57*
4. Designate trails for specific use	3.61 *1.33*	3.82 *0.90*	3.54[**] *1.47*	3.56 *0.90*	3.62 *1.72*	3.82 *0.90*	3.56[*] *0.90*	3.54 *1.47*	3.62 *1.72*
5. Develop new paved roads	3.19 *1.43*	3.21 *1.02*	3.35 *1.54*	2.89 *0.91*	3.01 *1.80*	3.21 *1.02*	2.89[*] *0.91*	3.35 *1.54*	3.01[**] *1.80*

continued on next page

Table G4. *Continued.*

Survey item number and statement	Full sample	East Non-metro	East Metro	West Non-metro	West Metro	Non-metro East	Non-metro West	Metro East	Metro West
17. Expand commercial recreation	3.45 *1.24*	3.31 *0.99*	3.42 *1.35*	3.34 *0.82*	3.58* *1.49*	3.31 *0.99*	3.34 *0.82*	3.42 *1.35*	3.58 *1.49*
19. Develop volunteer programs to improve facilities (trails, etc.)	3.79 *1.21*	3.93 *0.86*	3.83 *1.28*	3.70 *0.96*	3.68 *1.53*	3.93 *0.86*	3.70 *0.96*	3.83 *1.28*	3.68 *1.53*
20. Inform public about recreation concerns (safety, trail etiquette, etc.)	3.89 *1.31*	4.01 *0.95*	3.92 *1.36*	3.75 *0.76*	3.84 *1.73*	4.01 *0.95*	3.75* *0.76*	3.92 *1.36*	3.84 *1.73*
28. Collect an entry fee to support National Forests and Grasslands	3.61 *1.36*	3.73 *0.89*	3.56 *1.55*	3.40 *1.03*	3.66* *1.63*	3.73 *0.89*	3.40* *1.03*	3.56 *1.55*	3.66 *1.63*

	Values with respect to forests and grasslands *(1= strongly disagree, 5= strongly agree)*								
Survey item number and statement	Full sample	East Non-metro	East Metro	West Non-metro	West Metro	Non-metro East	Non-metro West	Metro East	Metro West
9. I would pay $5 more to use public lands for recreation	3.49 *1.60*	3.31 *1.24*	3.56** *1.70*	3.24 *1.10*	3.49* *1.98*	3.31 *1.24*	3.24 *1.10*	3.56 *1.70*	3.49 *1.98*

[a] Standard deviation

*, **, *** Mean difference is significant at   = 0.05, 0.01, 0.001

Strategic Plan Objective 2.b. Improve the capability of wilderness and protected areas to sustain a desired range of benefits and values (see table G5).

Table G5. Strategic Plan Objective 2.b.: Improve the capability of wilderness and protected areas to sustain a desired range of benefits and values. Survey items are from the VOBA module of the National Survey on Recreation and the Environment.

	Objectives for the management of forests and grasslands *(1= not at all important, 5= very important)*								
Survey item number and statement	Full sample	East Non-metro	East Metro	West Non-metro	West Metro	Non-metro East	Non-metro West	Metro East	Metro West
6. Designate more wilderness to stop development & motorized access	3.84 *1.41[a]*	3.70 *1.05*	3.90* *1.51*	3.28 *1.14*	3.89*** *1.66*	3.70 *1.05*	3.28** *1.14*	3.90 *1.51*	3.89 *1.66*
8. Preserve natural resources through policies such as no timber, no mining	4.22 *1.23*	3.99 *1.04*	4.29*** *1.26*	3.83 *1.05*	4.29*** *1.46*	3.99 *1.04*	3.83 *1.05*	4.29 *1.26*	4.29 *1.46*
9. Protecting ecosystems and wildlife habitat	4.58 *0.92*	4.49 *0.76*	4.61* *0.98*	4.23 *0.82*	4.62*** *1.02*	4.49 *0.76*	4.23** *0.82*	4.61 *0.98*	4.62 *1.02*
10. Preserve wilderness experience	4.15 *1.28*	3.96 *1.01*	4.25*** *1.31*	4.21 *0.77*	4.08 *1.65*	3.96 *1.01*	4.21* *0.77*	4.25 *1.31*	4.08* *1.65*

continued on next page

Table G5. *Continued.*

Survey item number and statement	Full sample	East		West		Non-metro		Metro	
		Non-metro	Metro	Non-metro	Metro	East	West	East	West

Beliefs
about the role of the USDA Forest Service
(1= strongly disagree, 5= strongly agree)

Survey item number and statement	Full sample	East		West		Non-metro		Metro	
		Non-metro	Metro	Non-metro	Metro	East	West	East	West
6. Designate more wilderness to stop development & motorized access	3.66 *1.46*	3.58 *1.06*	3.84[*] *1.54*	3.09 *1.22*	3.54[*] *1.75*	3.58 *1.06*	3.09[**] *1.22*	3.84 *1.54*	3.54[*] *1.75*
8. Preserve natural resources through policies such as no timber, no mining	4.21 *1.27*	4.05 *1.01*	4.29[*] *1.30*	3.83 *1.06*	4.22[*] *1.57*	4.05 *1.01*	3.83 *1.06*	4.29 *1.30*	4.22 *1.57*
9. Protecting ecosystems and wildlife habitat	4.53 *0.98*	4.41 *0.78*	4.55 *1.08*	4.40 *0.69*	4.56 *1.14*	4.41 *0.78*	4.40 *0.69*	4.55 *1.08*	4.56 *1.14*
10. Preserve wilderness experience	4.22 *1.14*	4.22 *0.86*	4.25 *1.23*	3.83 *0.97*	4.24[**] *1.33*	4.22 *0.86*	3.83[***] *0.97*	4.25 *1.23*	4.24 *1.33*

Attitudes
about the performance of the Forest Service
(1= very unfavorable, 5= very favorable)

Survey item number and statement	Full sample	East		West		Non-metro		Metro	
		Non-metro	Metro	Non-metro	Metro	East	West	East	West
6. Designate more wilderness to stop development & motorized access	3.45 *1.34*	3.34 *1.02*	3.35 *1.46*	3.24 *0.95*	3.65[**] *1.59*	3.34 *1.02*	3.24 *0.95*	3.35 *1.46*	3.65[**] *1.59*
8. Preserve natural resources through policies such as no timber, no mining	3.65 *1.31*	3.61 *1.03*	3.67 *1.38*	3.32 *0.90*	3.69[**] *1.63*	3.61 *1.03*	3.32[*] *0.90*	3.67 *1.38*	3.69 *1.63*
9. Protecting ecosystems and wildlife habitat	3.90 *1.16*	4.06 *0.80*	3.89[*] *1.28*	3.97 *0.83*	3.82 *1.42*	4.06 *0.80*	3.97 *0.83*	3.89 *1.28*	3.82 *1.42*
10. Preserve wilderness experience	3.88 *1.10*	3.92 *0.80*	3.87 *1.18*	3.83 *0.83*	3.90 *1.36*	3.92 *0.80*	3.83 *0.83*	3.87 *1.18*	3.90 *1.36*

[a] Standard deviation

[*], [**], [***] Mean difference is significant at = 0.05, 0.01, 0.001

Strategic Plan Objective 2.c. Improve the capability of the Nation's forests and grasslands to provide desired sustainable levels of uses, values, products, and services (see table G6).

Table G6. Strategic Plan Objective 2.c.: Improve the capability of the Nation's forests and rangelands to provide desired sustainable levels of uses, values, products, and services. Survey items are from the VOBA module of the National Survey on Recreation and the Environment.

		Objectives for the management of forests and grasslands (1= not at all important, 5= very important)							
Survey item number and statement	Full sample	East		West		Non-metro		Metro	
		Non-metro	Metro	Non-metro	Metro	East	West	East	West
11. Preserve cultural uses	3.82 *1.38[a]*	3.78 *0.97*	3.85 *1.51*	3.69 *1.06*	3.82 *1.68*	3.78 *0.97*	3.69 *1.06*	3.85 *1.51*	3.82 *1.68*
12. Provide natural resources to dependent communities	3.60 *1.39*	3.65 *1.03*	3.66 *1.53*	3.80 *0.89*	3.44** *1.70*	3.65 *1.03*	3.80 *0.89*	3.66 *1.53*	3.44* *1.70*
15. Make the permitting process easier for established uses	2.89 *1.55*	2.90 *1.14*	2.82 *1.69*	3.08 *1.06*	2.96 *1.91*	2.90 *1.14*	3.08 *1.06*	2.82 *1.69*	2.96 *1.91*
16. Develop a national policy for natural resource development	4.26 *1.23*	4.05 *0.97*	4.32** *1.31*	3.71 *0.98*	4.31*** *1.40*	4.05 *0.97*	3.71** *0.98*	4.32 *1.31*	4.31 *1.40*
18. Develop volunteer programs to improve land (tree planting, etc.)	4.60 *0.86*	4.47 *0.72*	4.63** *0.88*	4.42 *0.77*	4.65* *0.98*	4.47 *0.72*	4.42 *0.77*	4.63 *0.88*	4.65 *0.98*
25. Allow for diverse uses	4.07 *1.18*	4.00 *0.93*	4.16* *1.20*	4.11 *0.87*	3.95 *1.48*	4.00 *0.93*	4.11 *0.87*	4.16 *1.20*	3.95** *1.48*
27. Increase total number of acres in National Forest and Grassland system	3.81 *1.42*	3.64 *0.97*	3.97*** *1.49*	3.19 *1.07*	3.75*** *1.82*	3.64 *0.97*	3.19** *1.07*	3.97 *1.49*	3.75* *1.82*

		Beliefs about the role of the USDA Forest Service (1= strongly disagree, 5= strongly agree)							
Survey item number and statement	Full sample	East		West		Non-metro		Metro	
		Non-metro	Metro	Non-metro	Metro	East	West	East	West
11. Preserve cultural uses	3.75 *1.36*	3.80 *1.03*	3.81 *1.39*	3.21 *1.05*	3.73** *1.78*	3.80 *1.03*	3.21*** *1.05*	3.81 *1.39*	3.73 *1.78*
12. Provide natural resources to dependent communities	3.29 *1.39*	3.26 *1.02*	3.38 *1.54*	3.43 *0.93*	3.13 *1.64*	3.26 *1.02*	3.43 *0.93*	3.38 *1.54*	3.13* *1.64*
15. Make the permitting process easier for established uses	2.90 *1.59*	2.94 *1.17*	2.73 *1.70*	2.94 *1.09*	3.15 *2.02*	2.94 *1.17*	2.94 *1.09*	2.73 *1.70*	3.15* *2.02*
16. Develop a national policy for natural resource development	4.21 *1.19*	4.08 *1.05*	4.27* *1.21*	4.02 *0.81*	4.20 *1.46*	4.08 *1.05*	4.02 *0.81*	4.27 *1.21*	4.20 *1.46*
18. Develop volunteer programs to improve land (tree planting, etc.)	4.46 *1.03*	4.48 *0.77*	4.50 *1.02*	4.20 *0.84*	4.42 *1.38*	4.48 *0.77*	4.20* *0.84*	4.50 *1.02*	4.42 *1.38*

continued on next page

Table G6. *Continued.*

Survey item	Full sample	East Non-metro	East Metro	West Non-metro	West Metro	Non-metro East	Non-metro West	Metro East	Metro West
25. Allow for diverse uses	4.02 *1.14*	4.03 *0.90*	4.03 *1.25*	4.17 *0.74*	3.96 *1.37*	4.03 *0.90*	4.17 *0.74*	4.03 *1.25*	3.96 *1.37*
27. Increase total number of acres in National Forest and Grassland system	3.95 *1.47*	3.81 *1.02*	3.90 *1.54*	3.32 *1.05*	4.15*** *1.85*	3.81 *1.02*	3.32** *1.05*	3.90 *1.54*	4.15* *1.85*

	Attitudes about the performance of the Forest Service *(1= very unfavorable, 5= very favorable)*								
Survey item number and statement	Full sample	East		West		Non-metro		Metro	
		Non-metro	Metro	Non-metro	Metro	East	West	East	West
11. Preserve cultural uses	3.39 *1.37*	3.51 *0.97*	3.24** *1.63*	3.55 *0.95*	3.56 *1.45*	3.51 *0.97*	3.55 *0.95*	3.24 *1.63*	3.56*** *1.45*
12. Provide natural resources to dependent communities	3.45 *1.30*	3.46 *0.93*	3.57 *1.41*	3.04 *0.98*	3.35* *1.55*	3.46 *0.93*	3.04*** *0.98*	3.57 *1.41*	3.35* *1.55*
15. Make the permitting process easier for established uses	3.05 *1.43*	3.16 *1.09*	3.05 *1.55*	3.24 *0.97*	2.95* *1.84*	3.16 *1.09*	3.24 *0.97*	3.05 *1.55*	2.95 *1.84*
16. Develop a national policy for natural resource development	3.51 *1.26*	3.61 *0.91*	3.53 *1.41*	3.17 *0.92*	3.50** *1.51*	3.61 *0.91*	3.17*** *0.92*	3.53 *1.41*	3.50 *1.51*
18. Develop volunteer programs to improve land (tree planting, etc.)	3.85 *1.25*	3.76 *0.97*	3.91 *1.26*	3.78 *0.91*	3.81 *1.67*	3.76 *0.97*	3.78 *0.91*	3.91 *1.26*	3.81 *1.67*
25. Allow for diverse uses	3.73 *1.18*	3.72 *0.88*	3.71 *1.15*	3.76 *0.82*	3.76 *1.68*	3.72 *0.88*	3.76 *0.82*	3.71 *1.15*	3.76 *1.68*
27. Increase total number of acres in National Forest and Grassland system	3.52 *1.45*	3.51 *0.91*	3.42 *1.42*	3.68 *0.92*	3.62 *2.11*	3.51 *0.91*	3.68 *0.92*	3.42 *1.42*	3.62 *2.11*

[a] Standard deviation

*, **, *** Mean difference is significant at = 0.05, 0.01, 0.001

Strategic Plan Objective 2.d. Increase accessibility to a diversity of people and members of underserved and low-income populations to the full range of uses, values, products, and services (see table G7).

Table G7. Strategic Plan Objective 2.d.: Increase accessibility to a diversity of people and members of underserved and low-income populations to the full range of uses, values, products, and services. Survey items are from the VOBA module of the National Survey on Recreation and the Environment.

	Objectives for the management of forests and grasslands *(1= not at all important, 5= very important)*								
Survey item number and statement	Full sample	East		West		Non-metro		Metro	
		Non-metro	Metro	Non-metro	Metro	East	West	East	West
11. Preserve cultural uses	3.82 *1.38[a]*	3.78 *0.97*	3.85 *1.51*	3.69 *1.06*	3.82 *1.68*	3.78 *0.97*	3.69 *1.06*	3.85 *1.51*	3.82 *1.68*
12. Provide natural resources to dependent communities	3.60 *1.39*	3.65 *1.03*	3.66 *1.53*	3.80 *0.89*	3.44** *1.70*	3.65 *1.03*	3.80 *0.89*	3.66 *1.53*	3.44* *1.70*

continued on next page

Table G7. *Continued.*

Survey item number and statement	Full sample	East Non-metro	East Metro	West Non-metro	West Metro	Non-metro East	Non-metro West	Metro East	Metro West
15. Make the permitting process easier for established uses	2.89 *1.55*	2.90 *1.14*	2.82 *1.69*	3.08 *1.06*	2.96 *1.91*	2.90 *1.14*	3.08 *1.06*	2.82 *1.69*	2.96 *1.91*
16. Develop a national policy for natural resource development	4.26 *1.23*	4.05 *0.97*	4.32** *1.31*	3.71 *0.98*	4.31*** *1.40*	4.05 *0.97*	3.71** *0.98*	4.32 *1.31*	4.31 *1.40*
26. Make management decisions at local level (rather than national)	3.93 *1.22*	4.07 *0.85*	3.84** *1.40*	3.91 *0.84*	3.99 *1.40*	4.07 *0.85*	3.91 *0.84*	3.84 *1.40*	3.99 *1.40*

Beliefs
about the role of the USDA Forest Service
(1= strongly disagree, 5= strongly agree)

Survey item number and statement	Full sample	East Non-metro	East Metro	West Non-metro	West Metro	Non-metro East	Non-metro West	Metro East	Metro West
11. Preserve cultural uses	3.75 *1.36*	3.80 *1.03*	3.81 *1.39*	3.21 *1.05*	3.73** *1.78*	3.80 *1.03*	3.21*** *1.05*	3.81 *1.39*	3.73 *1.78*
12. Provide natural resources to dependent communities	3.29 *1.39*	3.26 *1.02*	3.38 *1.54*	3.43 *0.93*	3.13 *1.64*	3.26 *1.02*	3.43 *0.93*	3.38 *1.54*	3.13* *1.64*
15. Make the permitting process easier for established uses	2.90 *1.59*	2.94 *1.17*	2.73 *1.70*	2.94 *1.09*	3.15 *2.02*	2.94 *1.17*	2.94 *1.09*	2.73 *1.70*	3.15*** *2.02*
16. Develop a national policy for natural resource development	4.21 *1.19*	4.08 *1.05*	4.27* *1.21*	4.02 *0.81*	4.20 *1.46*	4.08 *1.05*	4.02 *0.81*	4.27 *1.21*	4.20 *1.46*
26. Make management decisions at local level (rather than national)	3.88 *1.34*	4.04 *0.94*	3.81* *1.44*	4.41 *0.80*	3.83** *1.69*	4.04 *0.94*	4.41* *0.80*	3.81 *1.44*	3.83 *1.69*

Attitudes
about the performance of the Forest Service
(1= very unfavorable, 5= very favorable)

Survey item number and statement	Full sample	East Non-metro	East Metro	West Non-metro	West Metro	Non-metro East	Non-metro West	Metro East	Metro West
11. Preserve cultural uses	3.39 *1.37*	3.51 *0.97*	3.24** *1.63*	3.55 *0.95*	3.56 *1.45*	3.51 *0.97*	3.55 *0.95*	3.24 *1.63*	3.56*** *1.45*
12. Provide natural resources to dependent communities	3.45 *1.30*	3.46 *0.93*	3.57 *1.41*	3.04 *0.98*	3.35* *1.55*	3.46 *0.93*	3.04*** *0.98*	3.57 *1.41*	3.35* *1.55*
15. Make the permitting process easier for established uses	3.05 *1.43*	3.16 *1.09*	3.05 *1.55*	3.24 *0.97*	2.95* *1.84*	3.16 *1.09*	3.24 *0.97*	3.05 *1.55*	2.95 *1.84*
16. Develop a national policy for natural resource development	3.51 *1.26*	3.61 *0.91*	3.53 *1.41*	3.17 *0.92*	3.50** *1.51*	3.61 *0.91*	3.17*** *0.92*	3.53 *1.41*	3.50 *1.51*
26. Make management decisions at local level (rather than national)	3.49 *1.32*	3.63 *0.98*	3.42* *1.39*	3.56 *0.94*	3.52 *1.68*	3.63 *0.98*	3.56 *0.94*	3.42 *1.39*	3.52 *1.68*

[a] Standard deviation

*, **, *** Mean difference is significant at = 0.05, 0.01, 0.001

Strategic Plan Objective 2.e. Improve delivery of services to urban communities (see table G8).

Table G8. Strategic Plan Objective 2.e.: Improve delivery of services to urban communities. Survey items are from the VOBA module of the National Survey on Recreation and the Environment.

		Objectives for the management of forests and grasslands *(1= not at all important, 5= very important)*							
		East		West		Non-metro		Metro	
Survey item number and statement	**Full sample**	Non-metro	Metro	Non-metro	Metro	East	West	East	West
1. Expand off-highway motorized access	2.41 *1.49[a]*	2.48 *1.11*	2.28* *1.55*	2.43 *0.98*	2.57 *1.97*	2.48 *1.11*	2.43 *0.98*	2.28 *1.55*	2.57** *1.97*
2. Trails for motorized vehicles	2.82 *1.47*	3.06 *1.11*	2.79** *1.60*	2.92 *0.98*	2.76 *1.75*	3.06 *1.11*	2.92 *0.98*	2.79 *1.60*	2.76 *1.75*
3. Trails for non-motorized recreation	3.75 *1.33*	3.56 *0.92*	3.80** *1.43*	3.27 *0.89*	3.82*** *1.67*	3.56 *0.92*	3.27* *0.89*	3.80 *1.43*	3.82 *1.67*
4. Designate trails for specific use	3.59 *1.43*	3.72 *0.98*	3.65 *1.63*	3.83 *1.01*	3.36** *1.63*	3.72 *0.98*	3.83 *1.01*	3.65 *1.63*	3.36** *1.63*
5. Develop new paved roads	2.62 *1.49*	2.59 *1.14*	2.57 *1.64*	2.45 *1.05*	2.76* *1.80*	2.59 *1.14*	2.45 *1.05*	2.57 *1.64*	2.76 *1.80*
10. Preserve "wilderness" experience	4.15 *1.28*	3.96 *1.01*	4.25*** *1.31*	4.21 *0.77*	4.08 *1.65*	3.96 *1.01*	4.21* *0.77*	4.25 *1.31*	4.08* *1.65*
17. Expand commercial recreation	3.04 *1.45*	3.02 *1.00*	2.87 *1.49*	3.18 *1.03*	3.31 *1.95*	3.02 *1.00*	3.18 *1.03*	2.87 *1.49*	3.31*** *1.95*

		Beliefs about the role of the USDA Forest Service *(1= strongly disagree, 5= strongly agree)*							
		East		West		Non-metro		Metro	
Survey item number and statement	**Full sample**	Non-metro	Metro	Non-metro	Metro	East	West	East	West
1. Expand off-highway motorized access	2.52 *1.50*	2.67 *1.12*	2.50 *1.61*	2.70 *1.17*	2.47 *1.76*	2.67 *1.12*	2.70 *1.17*	2.50 *1.61*	2.47 *1.76*
2. Trails for motorized vehicles	2.98 *1.51*	2.95 *1.16*	2.86 *1.71*	2.84 *1.04*	3.22 *1.72*	2.95 *1.16*	2.84 *1.04*	2.86 *1.71*	3.22** *1.72*
3. Trails for non-motorized recreation	3.71 *1.37*	3.45 *1.07*	3.75* *1.49*	3.19 *0.96*	3.82** *1.59*	3.45 *1.07*	3.19 *0.96*	3.75 *1.49*	3.82 *1.59*
4. Designate trails for specific use	3.94 *1.25*	4.11 *0.93*	3.92 *1.34*	4.00 *0.92*	3.88 *1.53*	4.11 *0.93*	4.00 *0.92*	3.92 *1.34*	3.88 *1.53*
5. Develop new paved roads	2.70 *1.57*	2.65 *1.05*	2.53 *1.67*	2.24 *0.95*	3.02** *2.05*	2.65 *1.05*	2.24* *0.95*	2.53 *1.67*	3.02*** *2.05*
10. Preserve "wilderness" experience	4.22 *1.14*	4.22 *0.86*	4.25 *1.23*	3.83 *0.97*	4.24** *1.33*	4.22 *0.86*	3.83*** *0.97*	4.25 *1.23*	4.24 *1.33*
17. Expand commercial recreation	3.25 *1.53*	3.20 *1.05*	3.15 *1.63*	2.87 *1.00*	3.48** *1.97*	3.20 *1.05*	2.87* *1.00*	3.15 *1.63*	3.48** *1.97*

continued on next page

Table G8. *Continued.*

Attitudes
about the performance of the Forest Service
(1= very unfavorable, 5= very favorable)

Survey item number and statement	Full sample	East		West		Non-metro		Metro	
		Non-metro	Metro	Non-metro	Metro	East	West	East	West
1. Expand off-highway motorized access	2.97 *1.45*	3.06 *1.03*	2.95 *1.61*	2.97 *1.00*	2.95 *1.82*	3.06 *1.03*	2.97 *1.00*	2.95 *1.61*	2.95 *1.82*
2. Trails for motorized vehicles	3.25 *1.38*	3.24 *0.94*	3.24 *1.51*	3.24 *0.96*	3.29 *1.71*	3.24 *0.94*	3.24 *0.96*	3.24 *1.51*	3.29 *1.71*
3. Trails for non-motorized recreation	3.59 *1.27*	3.54 *0.93*	3.53 *1.36*	3.34 *1.09*	3.74** *1.57*	3.54 *0.93*	3.34 *1.09*	3.53 *1.36*	3.74* *1.57*
4. Designate trails for specific use	3.61 *1.33*	3.82 *0.90*	3.54** *1.47*	3.56 *0.90*	3.62 *1.72*	3.82 *0.90*	3.56* *0.90*	3.54 *1.47*	3.62 *1.72*
5. Develop new paved roads	3.19 *1.43*	3.21 *1.02*	3.35 *1.54*	2.89 *0.91*	3.01 *1.80*	3.21 *1.02*	2.89* *0.91*	3.35 *1.54*	3.01** *1.80*
10. Preserve "wilderness" experience	3.88 *1.10*	3.92 *0.80*	3.87 *1.18*	3.83 *0.83*	3.90 *1.36*	3.92 *0.80*	3.83 *0.83*	3.87 *1.18*	3.90 *1.36*
17. Expand commercial recreation	3.45 *1.24*	3.31 *0.99*	3.42 *1.35*	3.34 *0.82*	3.58* *1.49*	3.31 *0.99*	3.34 *0.82*	3.42 *1.35*	3.58 *1.49*

[a] Standard deviation
*, **, *** Mean difference is significant at   = 0.05, 0.01, 0.001

Goal 3: Scientific and Technical Assistance: Develop and use the best scientific information available to deliver technical and community assistance and to support ecological, economic, and social sustainability.

Strategic Plan Objective 3.a. Better assist in building the capacity of Tribal governments, rural communities, and private landowners to adapt to economic, environmental, and social change related to natural resources (see table G9).

Table G9. Strategic Plan Objective 3.a.: Better assist in building the capacity of Tribal governments, rural communities, and private landowners to adapt to economic, environmental, and social change related to natural resources. Survey items are from the VOBA module of the National Survey on Recreation and the Environment.

Objectives
for the management of forests and grasslands
(1= not at all important, 5= very important)

Survey item number and statement	Full sample	East		West		Non-metro		Metro	
		Non-metro	Metro	Non-metro	Metro	East	West	East	West
11. Preserve cultural uses	3.82 *1.38[a]*	3.78 *0.97*	3.85 *1.51*	3.69 *1.06*	3.82 *1.68*	3.78 *0.97*	3.69 *1.06*	3.85 *1.51*	3.82 *1.68*
20. Inform public about recreation concerns (safety, trail etiquette, etc.)	4.53 *0.93*	4.47 *0.77*	4.58* *0.97*	4.35 *0.83*	4.51* *1.08*	4.47 *0.77*	4.35 *0.83*	4.58 *0.97*	4.51 *1.08*

continued on next page

Table G9. *Continued.*

Survey item number and statement	Full sample	East Non-metro	East Metro	West Non-metro	West Metro	Non-metro East	Non-metro West	Metro East	Metro West
21. Inform public on potential environmental impacts of uses	4.39 *1.02*	4.29 *0.82*	4.54*** *0.94*	3.86 *0.95*	4.31*** *1.37*	4.29 *0.82*	3.86*** *0.95*	4.54 *0.94*	4.31** *1.37*
22. Inform public on economic value from developing natural resources	4.02 *1.30*	3.93 *1.03*	4.10* *1.36*	3.90 *1.06*	3.95 *1.55*	3.93 *1.03*	3.90 *1.06*	4.10 *1.36*	3.95 *1.55*
23. Encourage collaboration between groups to share information	4.15 *1.20*	4.23 *0.80*	4.23 *1.23*	3.93 *0.85*	4.03 *1.64*	4.23 *0.80*	3.93** *0.85*	4.23 *1.23*	4.03* *1.64*
24. Using public advisory committees to advise on management issues	3.90 *1.20*	3.83 *0.94*	3.96 *1.23*	3.66 *0.79*	3.89* *1.56*	3.83 *0.94*	3.66 *0.79*	3.96 *1.23*	3.89 *1.56*
26. Make management decisions at local level (rather than national)	3.93 *1.22*	4.07 *0.85*	3.84** *1.40*	3.91 *0.84*	3.99 *1.40*	4.07 *0.85*	3.91 *0.84*	3.84 *1.40*	3.99 *1.40*

Beliefs
about the role of the USDA Forest Service
(1= strongly disagree, 5= strongly agree)

Survey item number and statement	Full sample	East Non-metro	East Metro	West Non-metro	West Metro	Non-metro East	Non-metro West	Metro East	Metro West
11. Preserve cultural uses	3.75 *1.36*	3.80 *1.03*	3.81 *1.39*	3.21 *1.05*	3.73** *1.78*	3.80 *1.03*	3.21 *1.05*	3.81*** *1.39*	3.73 *1.78*
20. Inform public about recreation concerns (safety, trail etiquette, etc.)	4.50 *0.93*	4.50 *0.71*	4.55 *1.00*	4.42 *0.72*	4.45 *1.12*	4.50 *0.71*	4.42 *0.72*	4.55 *1.00*	4.45 *1.12*
21. Inform public on potential environmental impacts of uses	4.48 *1.00*	4.32 *0.78*	4.48 *1.10*	4.46 *0.64*	4.54 *1.18*	4.32 *0.78*	4.46 *0.64*	4.48 *1.10*	4.54 *1.18*
22. Inform public on economic value from developing natural resources	4.08 *1.17*	4.20 *0.85*	4.11 *1.28*	3.93 *0.94*	3.99 *1.39*	4.20 *0.85*	3.93 *0.94*	4.11* *1.28*	3.99 *1.39*
23. Encourage collaboration between groups to share information	4.15 *1.17*	4.21 *0.82*	4.12 *1.30*	4.07 *0.84*	4.19 *1.42*	4.21 *0.82*	4.07 *0.84*	4.12 *1.30*	4.19 *1.42*
24. Using public advisory committees to advise on management issues	3.84 *1.25*	4.04 *0.85*	3.74* *1.41*	3.98 *0.91*	3.92 *1.44*	4.04 *0.85*	3.98 *0.91*	3.74 *1.41*	3.92 *1.44*
26. Make management decisions at local level (rather than national)	3.88 *1.34*	4.04 *0.94*	3.81* *1.44*	4.41 *0.80*	3.83** *1.69*	4.04 *0.94*	4.41 *0.80*	3.81* *1.44*	3.83 *1.69*

Attitudes
about the performance of the Forest Service
(1= very unfavorable, 5= very favorable)

Survey item number and statement	Full sample	East Non-metro	East Metro	West Non-metro	West Metro	Non-metro East	Non-metro West	Metro East	Metro West
11. Preserve cultural uses	3.39 *1.37*	3.51 *0.97*	3.24** *1.63*	3.55 *0.95*	3.56 *1.45*	3.51 *0.97*	3.55 *0.95*	3.24 *1.63*	3.56*** *1.45*

continued on next page

90

Table G9. *Continued.*

Survey item number and statement	Full sample	East Non-metro	East Metro	West Non-metro	West Metro	Non-metro East	Non-metro West	Metro East	Metro West
20. Inform public about recreation concerns (safety, trail etiquette, etc.)	3.89 *1.31*	4.01 *0.95*	3.92 *1.36*	3.75 *0.76*	3.84 *1.73*	4.01 *0.95*	3.75* *0.76*	3.92 *1.36*	3.84 *1.73*
21. Inform public on potential environmental impacts of uses	3.50 *1.33*	3.61 *0.97*	3.48 *1.39*	3.40 *0.90*	3.50 *1.77*	3.61 *0.97*	3.40 *0.90*	3.48 *1.39*	3.50 *1.77*
22. Inform public on economic value from developing natural resources	3.40 *1.38*	3.38 *1.00*	3.38 *1.47*	3.42 *1.07*	3.43 *1.72*	3.38 *1.00*	3.42 *1.07*	3.38 *1.47*	3.43 *1.72*
23. Encourage collaboration between groups to share information	3.72 *1.22*	3.67 *0.90*	3.65 *1.40*	3.76 *0.80*	3.83 *1.46*	3.67 *0.90*	3.76 *0.80*	3.65 *1.40*	3.83* *1.46*
24. Using public advisory committees to advise on management issues	3.36 *1.26*	3.39 *0.93*	3.35 *1.48*	3.34 *0.82*	3.35 *1.42*	3.39 *0.93*	3.34 *0.82*	3.35 *1.48*	3.35 *1.42*
26. Make management decisions at local level (rather than national)	3.49 *1.32*	3.63 *0.98*	3.42* *1.39*	3.56 *0.94*	3.52 *1.68*	3.63 *0.98*	3.56 *0.94*	3.42 *1.39*	3.52 *1.68*

[a] Standard deviation
*, **, *** Mean difference is significant at = 0.05, 0.01, 0.001

Strategic Plan Objective 3.b. Increase the effectiveness of scientific, developmental, and technical information delivered to domestic and international interests (see table G10).

Table G10. Strategic Plan Objective 3.b.: Increase the effectiveness of scientific, developmental, and technical information delivered to domestic and international interests. Survey items are from the VOBA module of the National Survey on Recreation and the Environment.

Survey item number and statement	Full sample	Objectives for the management of forests and grasslands (1= not at all important, 5= very important)							
		East Non-metro	East Metro	West Non-metro	West Metro	Non-metro East	Non-metro West	Metro East	Metro West
20. Inform public about recreation concerns (safety, trail etiquette, etc.)	4.53 *0.93[a]*	4.47 *0.77*	4.58* *0.97*	4.35 *0.83*	4.51* *1.08*	4.47 *0.77*	4.35 *0.83*	4.58 *0.97*	4.51 *1.08*
21. Inform public on potential environmental impacts of uses	4.39 *1.02*	4.29 *0.82*	4.54*** *0.94*	3.86 *0.95*	4.31*** *1.37*	4.29 *0.82*	3.86*** *0.95*	4.54 *0.94*	4.31** *1.37*
22. Inform public on economic value from developing natural resources	4.02 *1.30*	3.93 *1.03*	4.10* *1.36*	3.90 *1.06*	3.95 *1.55*	3.93 *1.03*	3.90 *1.06*	4.10 *1.36*	3.95 *1.55*
23. Encourage collaboration between groups to share information	4.15 *1.20*	4.23 *0.80*	4.23 *1.23*	3.93 *0.85*	4.03 *1.64*	4.23 *0.80*	3.93** *0.85*	4.23 *1.23*	4.03* *1.64*

continued on next page

Beliefs

about the role of the USDA Forest Service

(1= strongly disagree, 5= strongly agree)

Survey item number and statement	Full sample	East		West		Non-metro		Metro	
		Non-metro	Metro	Non-metro	Metro	East	West	East	West
20. Inform public about recreation concerns (safety, trail etiquette, etc.)	4.50 *0.93*	4.50 *0.71*	4.55 *1.00*	4.42 *0.72*	4.45 *1.12*	4.50 *0.71*	4.42 *0.72*	4.55 *1.00*	4.45 *1.12*
21. Inform public on potential environmental impacts of uses	4.48 *1.00*	4.32 *0.78*	4.48 *1.10*	4.46 *0.64*	4.54 *1.18*	4.32 *0.78*	4.46 *0.64*	4.48 *1.10*	4.54 *1.18*
22. Inform public on economic value from developing natural resources	4.08 *1.17*	4.20 *0.85*	4.11 *1.28*	3.93 *0.94*	3.99 *1.39*	4.20 *0.85*	3.93* *0.94*	4.11 *1.28*	3.99 *1.39*
23. Encourage collaboration between groups to share information	4.15 *1.17*	4.21 *0.82*	4.12 *1.30*	4.07 *0.84*	4.19 *1.42*	4.21 *0.82*	4.07 *0.84*	4.12 *1.30*	4.19 *1.42*

Attitudes

about the performance of the Forest Service

(1= very unfavorable, 5= very favorable)

Survey item number and statement	Full sample	East		West		Non-metro		Metro	
		Non-metro	Metro	Non-metro	Metro	East	West	East	West
20. Inform public about recreation concerns (safety, trail etiquette, etc.)	3.89 *1.31*	4.01 *0.95*	3.92 *1.36*	3.75 *0.76*	3.84 *1.73*	4.01 *0.95*	3.75* *0.76*	3.92 *1.36*	3.84 *1.73*
21. Inform public on potential environmental impacts of uses	3.50 *1.33*	3.61 *0.97*	3.48 *1.39*	3.40 *0.90*	3.50 *1.77*	3.61 *0.97*	3.40 *0.90*	3.48 *1.39*	3.50 *1.77*
22. Inform public on economic value from developing natural resources	3.40 *1.38*	3.38 *1.00*	3.38 *1.47*	3.42 *1.07*	3.43 *1.72*	3.38 *1.00*	3.42 *1.07*	3.38 *1.47*	3.43 *1.72*
23. Encourage collaboration between groups to share information	3.72 *1.22*	3.67 *0.90*	3.65 *1.40*	3.76 *0.80*	3.83 *1.46*	3.67 *0.90*	3.76 *0.80*	3.65 *1.40*	3.83* *1.46*

[a] Standard deviation

*, **, *** Mean difference is significant at = 0.05, 0.01, 0.001

Strategic Plan Objective 3.c. Improve the knowledge base provided through research, inventory, and monitoring to enhance scientific understanding of ecosystems, including human uses, and to support decision-making and sustainable management of the Nation's forests and grasslands (see table G11).

Table G11. Strategic Plan Objective 3.c.: Improve the knowledge base provided through research, inventory, and monitoring to enhance scientific understanding of ecosystems, including human uses, and to support decision-making and sustainable management of the Nation's forests and grasslands. Survey items are from the VOBA module of the National Survey on Recreation and the Environment.

Survey item number and statement	Full sample	**Objectives** for the management of forests and grasslands (1= not at all important, 5= very important)							
		East		West		Non-metro		Metro	
		Non-metro	Metro	Non-metro	Metro	East	West	East	West
20. Inform public about recreation concerns (safety, trail etiquette, etc.)	4.53 *0.93*[a]	4.47 *0.77*	4.58* *0.97*	4.35 *0.83*	4.51* *1.08*	4.47 *0.77*	4.35 *0.83*	4.58 *0.97*	4.51 *1.08*
21. Inform public on potential environmental impacts of uses	4.39 *1.02*	4.29 *0.82*	4.54*** *0.94*	3.86 *0.95*	4.31*** *1.37*	4.29 *0.82*	3.86*** *0.95*	4.54 *0.94*	4.31** *1.37*
22. Inform public on economic value from developing natural resources	4.02 *1.30*	3.93 *1.03*	4.10* *1.36*	3.90 *1.06*	3.95 *1.55*	3.93 *1.03*	3.90 *1.06*	4.10 *1.36*	3.95 *1.55*
23. Encourage collaboration between groups to share information	4.15 *1.20*	4.23 *0.80*	4.23 *1.23*	3.93 *0.85*	4.03 *1.64*	4.23 *0.80*	3.93** *0.85*	4.23 *1.23*	4.03* *1.64*

Survey item number and statement	Full sample	**Beliefs** about the role of the USDA Forest Service (1= strongly disagree, 5= strongly agree)							
		East		West		Non-metro		Metro	
		Non-metro	Metro	Non-metro	Metro	East	West	East	West
20. Inform public about recreation concerns (safety, trail etiquette, etc.)	4.50 *0.93*	4.50 *0.71*	4.55 *1.00*	4.42 *0.72*	4.45 *1.12*	4.50 *0.71*	4.42 *0.72*	4.55 *1.00*	4.45 *1.12*
21. Inform public on potential environmental impacts of uses	4.48 *1.00*	4.32 *0.78*	4.48 *1.10*	4.46 *0.64*	4.54 *1.18*	4.32 *0.78*	4.46 *0.64*	4.48 *1.10*	4.54 *1.18*
22. Inform public on economic value from developing natural resources	4.08 *1.17*	4.20 *0.85*	4.11 *1.28*	3.93 *0.94*	3.99 *1.39*	4.20 *0.85*	3.93* *0.94*	4.11 *1.28*	3.99 *1.39*
23. Encourage collaboration between groups to share information	4.15 *1.17*	4.21 *0.82*	4.12 *1.30*	4.07 *0.84*	4.19 *1.42*	4.21 *0.82*	4.07 *0.84*	4.12 *1.30*	4.19 *1.42*

Survey item number and statement	Full sample	**Attitudes** about the performance of the Forest Service (1= very unfavorable, 5= very favorable)							
		East		West		Non-metro		Metro	
		Non-metro	Metro	Non-metro	Metro	East	West	East	West
20. Inform public about recreation concerns (safety, trail etiquette, etc.)	3.89 *1.31*	4.01 *0.95*	3.92 *1.36*	3.75 *0.76*	3.84 *1.73*	4.01 *0.95*	3.75* *0.76*	3.92 *1.36*	3.84 *1.73*
21. Inform public on potential environmental impacts of uses	3.50 *1.33*	3.61 *0.97*	3.48 *1.39*	3.40 *0.90*	3.50 *1.77*	3.61 *0.97*	3.40 *0.90*	3.48 *1.39*	3.50 *1.77*
22. Inform public on economic value from developing natural resources	3.40 *1.38*	3.38 *1.00*	3.38 *1.47*	3.42 *1.07*	3.43 *1.72*	3.38 *1.00*	3.42 *1.07*	3.38 *1.47*	3.43 *1.72*

continued on next page

23. Encourage collaboration between groups to share information	3.72 *1.22*	3.67 *0.90*	3.65 *1.40*	3.76 *0.80*	3.83 *1.46*	3.67 *0.90*	3.76 *0.80*	3.65 *1.40*	3.83[*] *1.46*

[a] Standard deviation

[*], [**], [***] Mean difference is significant at = 0.05, 0.01, 0.001

Strategic Plan Objective 3.d. Broaden the participation of less traditional research groups in research and technical assistance programs (see table G12).

Table G12. Strategic Plan Objective 3.d.: Broaden the participation of less traditional research groups in research and technical assistance programs. Survey items are from the VOBA module of the National Survey on Recreation and the Environment.

		Objectives for the management of forests and grasslands (*1= not at all important, 5= very important*)							
Survey item number and statement	Full sample	East		West		Non-metro		Metro	
		Non-metro	Metro	Non-metro	Metro	East	West	East	West
23. Encourage collaboration between groups to share information	4.15 *1.20[a]*	4.23 *0.80*	4.23 *1.23*	3.93 *0.85*	4.03 *1.64*	4.23 *0.80*	3.93[**] *0.85*	4.23 *1.23*	4.03[*] *1.64*

		Beliefs about the role of the USDA Forest Service (*1= strongly disagree, 5= strongly agree*)							
Survey item number and statement	Full sample	East		West		Non-metro		Metro	
		Non-metro	Metro	Non-metro	Metro	East	West	East	West
23. Encourage collaboration between groups to share information	4.15 *1.17*	4.21 *0.82*	4.12 *1.30*	4.07 *0.84*	4.19 *1.42*	4.21 *0.82*	4.07 *0.84*	4.12 *1.30*	4.19 *1.42*

		Attitudes about the performance of the Forest Service (*1= very unfavorable, 5= very favorable*)							
Survey item number and statement	Full sample	East		West		Non-metro		Metro	
		Non-metro	Metro	Non-metro	Metro	East	West	East	West
23. Encourage collaboration between groups to share information	3.72 *1.22*	3.67 *0.90*	3.65 *1.40*	3.76 *0.80*	3.83 *1.46*	3.67 *0.90*	3.76 *0.80*	3.65 *1.40*	3.83[*] *1.46*

[a] Standard deviation

[*], [**], [***] Mean difference is significant at = 0.05, 0.01, 0.001

Goal 4: Effective Public Service: Ensure the acquisition and use of an appropriate corporate infrastructure to enable the efficient delivery of a variety of uses.

Strategic Plan Objective 4.a. Improve financial management to achieve fiscal accountability (see table G13).

Table G13. Strategic Plan Objective 4.a.: Improve financial management to achieve fiscal accountability. Survey item is from the VOBA module of the National Survey on Recreation and the Environment.

Survey item number and statement	Full sample	Values with respect to forests and grasslands (1= strongly disagree, 5= strongly agree)							
		East		West		Non-metro		Metro	
		Non-metro	Metro	Non-metro	Metro	East	West	East	West
24. Federal government should subsidize development and leasing of public lands	2.32 1.58^a	2.33 _1.15_	2.28 _1.73_	2.09 _0.98_	2.42^{**} _1.95_	2.33 _1.15_	2.09^* _0.98_	2.28 _1.73_	2.42 _1.95_

[a] Standard deviation
*, **, *** Mean difference is significant at = 0.05, 0.01, 0.001

Strategic Plan Objective 4.b. Improve the safety and economy of USDA Forest Service roads, trails, facilities, and operations and provide greater security for the public and employees (see table G14).

Table G14. Strategic Plan Objective 4.b.: Improve the safety and economy of USDA Forest Service roads, trails, facilities, and operations and provide greater security for the public and employees. Survey items are from the VOBA module of the National Survey on Recreation and the Environment.

Survey item number and statement	Full sample	Objectives for the management of forests and grasslands (1= not at all important, 5= very important)							
		East		West		Non-metro		Metro	
		Non-metro	Metro	Non-metro	Metro	East	West	East	West
19. Develop volunteer programs to improve facilities (trails, etc.)	4.18 1.13^a	4.04 _0.87_	4.09 _1.31_	4.32 _0.66_	4.34 _1.24_	4.04 _0.87_	4.32^{**} _0.66_	4.09 _1.31_	4.34^{***} _1.24_
20. Inform public about recreation concerns (safety, trail etiquette, etc.)	4.53 _0.93_	4.47 _0.77_	4.58^* _0.97_	4.35 _0.83_	4.51^* _1.08_	4.47 _0.77_	4.35 _0.83_	4.58 _0.97_	4.51 _1.08_
29. Increasing law enforcement on National Forests and Grasslands	4.01 _1.21_	3.96 _0.97_	3.98 _1.30_	3.65 _0.88_	4.12^{***} _1.50_	3.96 _0.97_	3.65^* _0.88_	3.98 _1.30_	4.12 _1.50_

Survey item number and statement	Full sample	Beliefs about the role of the USDA Forest Service (1= strongly disagree, 5= strongly agree)							
		East		West		Non-metro		Metro	
		Non-metro	Metro	Non-metro	Metro	East	West	East	West
19. Develop volunteer programs to improve facilities (trails, etc.)	4.22 _1.09_	4.18 _0.85_	4.20 _1.18_	4.18 _0.82_	4.28 _1.31_	4.18 _0.85_	4.18 _0.82_	4.20 _1.18_	4.28 _1.31_
20. Inform public about recreation concerns (safety, trail etiquette, etc.)	4.50 _0.93_	4.50 _0.71_	4.55 _1.00_	4.42 _0.72_	4.45 _1.12_	4.50 _0.71_	4.42 _0.72_	4.55 _1.00_	4.45 _1.12_
29. Increasing law enforcement on National Forests and Grasslands	4.01 _1.26_	4.08 _0.91_	4.09 _1.31_	4.03 _0.83_	3.87 _1.68_	4.08 _0.91_	4.03 _0.83_	4.09 _1.31_	3.87^* _1.68_

continued on next page

Table G14. *Continued.*

Survey item number and statement	Full sample	**Attitudes** about the performance of the Forest Service *(1= very unfavorable, 5= very favorable)*								
		East		West		Non-metro		Metro		
		Non-metro	Metro	Non-metro	Metro	East	West	East	West	
19. Develop volunteer programs to improve facilities (trails, etc.)	3.79 *1.21*	3.93 *0.86*	3.83 *1.28*	3.70 *0.96*	3.68 *1.53*	3.93 *0.86*	3.70 *0.96*	3.83 *1.28*	3.68 *1.53*	
20. Inform public about recreation concerns (safety, trail etiquette, etc.)	3.89 *1.31*	4.01 *0.95*	3.92 *1.36*	3.75 *0.76*	3.84 *1.73*	4.01 *0.95*	3.75* *0.76*	3.92 *1.36*	3.84 *1.73*	
29. Increasing law enforcement on National Forests and Grasslands	3.85 *1.27*	3.85 *0.95*	3.89 *1.37*	3.48 *0.98*	3.85* *1.61*	3.85 *0.95*	3.48** *0.98*	3.89 *1.37*	3.85 *1.61*	

[a] Standard deviation
*, **, *** Mean difference is significant at α = 0.05, 0.01, 0.001

Strategic Plan Objective 4.f. Provide appropriate access to National Forest System lands and ensure nondiscrimination in the delivery of all USDA Forest Service programs (see table G15).

Table G15. Strategic Plan Objective 4.f.: Provide appropriate access to National Forest System lands and ensure nondiscrimination in the delivery of all USDA Forest Service programs. Survey items are from the VOBA module of the National Survey on Recreation and the Environment.

Survey item number and statement	Full sample	**Objectives** for the management of forests and grasslands *(1= not at all important, 5= very important)*								
		East		West		Non-metro		Metro	
		Non-metro	Metro	Non-metro	Metro	East	West	East	West
11. Preserve cultural uses	3.82 *1.38*[a]	3.78 *0.97*	3.85 *1.51*	3.69 *1.06*	3.82 *1.68*	3.78 *0.97*	3.69 *1.06*	3.85 *1.51*	3.82 *1.68*
12. Provide natural resources to dependent communities	3.60 *1.39*	3.65 *1.03*	3.66 *1.53*	3.80 *0.89*	3.44** *1.70*	3.65 *1.03*	3.80 *0.89*	3.66 *1.53*	3.44* *1.70*
15. Make the permitting process easier for established uses	2.89 *1.55*	2.90 *1.14*	2.82 *1.69*	3.08 *1.06*	2.96 *1.91*	2.90 *1.14*	3.08 *1.06*	2.82 *1.69*	2.96 *1.91*
17. Expand commercial recreation	3.04 *1.45*	3.02 *1.00*	2.87 *1.49*	3.18 *1.03*	3.31 *1.95*	3.02 *1.00*	3.18 *1.03*	2.87 *1.49*	3.31*** *1.95*
25. Allow for diverse uses	4.07 *1.18*	4.00 *0.93*	4.16* *1.20*	4.11 *0.87*	3.95 *1.48*	4.00 *0.93*	4.11 *0.87*	4.16 *1.20*	3.95** *1.48*

Survey item number and statement	Full sample	**Beliefs** about the role of the USDA Forest Service *(1= strongly disagree, 5= strongly agree)*								
		East		West		Non-metro		Metro	
		Non-metro	Metro	Non-metro	Metro	East	West	East	West
11. Preserve cultural uses	3.75 *1.36*	3.80 *1.03*	3.81 *1.39*	3.21 *1.05*	3.73** *1.78*	3.80 *1.03*	3.21*** *1.05*	3.81 *1.39*	3.73 *1.78*

continued on next page

Table G15. *Continued.*

Survey item number and statement	Full sample	East Non-metro	East Metro	West Non-metro	West Metro	Non-metro East	Non-metro West	Metro East	Metro West
12. Provide natural resources to dependent communities	3.29	3.26	3.38	3.43	3.13	3.26	3.43	3.38	3.13[*]
	1.39	*1.02*	*1.54*	*0.93*	*1.64*	*1.02*	*0.93*	*1.54*	*1.64*
15. Make the permitting process easier for established uses	2.90	2.94	2.73	2.94	3.15	2.94	2.94	2.73	3.15[***]
	1.59	*1.17*	*1.70*	*1.09*	*2.02*	*1.17*	*1.09*	*1.70*	*2.02*
17. Expand commercial recreation	3.25	3.20	3.15	2.87	3.48[**]	3.20	2.87[*]	3.15	3.48[**]
	1.53	*1.05*	*1.63*	*1.00*	*1.97*	*1.05*	*1.00*	*1.63*	*1.97*
25. Allow for diverse uses	4.02	4.03	4.03	4.17	3.96	4.03	4.17	4.03	3.96
	1.14	*0.90*	*1.25*	*0.74*	*1.37*	*0.90*	*0.74*	*1.25*	*1.37*

	Attitudes about the performance of the Forest Service *(1= very unfavorable, 5= very favorable)*								
Survey item number and statement	**Full sample**	**East**		**West**		**Non-metro**		**Metro**	
		Non-metro	Metro	Non-metro	Metro	East	West	East	West
11. Preserve cultural uses	3.39	3.51	3.24[**]	3.55	3.56	3.51	3.55	3.24	3.56[***]
	1.37	*0.97*	*1.63*	*0.95*	*1.45*	*0.97*	*0.95*	*1.63*	*1.45*
12. Provide natural resources to dependent communities	3.45	3.46	3.57	3.04	3.35[*]	3.46	3.04[***]	3.57	3.35[*]
	1.30	*0.93*	*1.41*	*0.98*	*1.55*	*0.93*	*0.98*	*1.41*	*1.55*
15. Make the permitting process easier for established uses	3.05	3.16	3.05	3.24	2.95[*]	3.16	3.24	3.05	2.95
	1.43	*1.09*	*1.55*	*0.97*	*1.84*	*1.09*	*0.97*	*1.55*	*1.84*
17. Expand commercial recreation	3.45	3.31	3.42	3.34	3.58[*]	3.31	3.34	3.42	3.58
	1.24	*0.99*	*1.35*	*0.82*	*1.49*	*0.99*	*0.82*	*1.35*	*1.49*
25. Allow for diverse uses	3.73	3.72	3.71	3.76	3.76	3.72	3.76	3.71	3.76
	1.18	*0.88*	*1.15*	*0.82*	*1.68*	*0.88*	*0.82*	*1.15*	*1.68*

[a] Standard deviation

[*], [**], [***] Mean difference is significant at α = 0.05, 0.01, 0.001

Appendix H. Weighted Means by Demographic Breakdowns Grouped by Public Lands Values

Socially Responsible Individual Values (see table H1).

Socially Responsible Management Values (see table H2).

Table H1. Socially Responsible Individual Values. Survey items are from the VOBA module of the National Survey on Recreation and the Environment.

Survey item number and statement	Full sample	Values with respect to forests and grasslands (1=strongly disagree, 5=strongly agree)							
		East		West		Non-metro		Metro	
		Non-metro	Metro	Non-metro	Metro	East	West	East	West
1. People should be more concerned about how pub lands are used	4.75 *0.72ª*	4.74 *0.57*	4.79 *0.70*	4.61 *0.55*	4.71 *1.01*	4.74 *0.57*	4.61[*] *0.55*	4.79 *0.70*	4.71 *1.01*
2. Natural resources must be preserved, even if some people must do without	4.14 *1.22*	3.83 *1.04*	4.23[***] *1.23*	4.22 *0.85*	4.11 *1.51*	3.83 *1.04*	4.22[***] *0.85*	4.23 *1.23*	4.11 *1.51*
3. Consumers should be interested in environmental consequences of purchases	4.47 *1.05*	4.63 *0.60*	4.46[**] *1.18*	4.27 *0.88*	4.46 *1.28*	4.63 *0.60*	4.27[***] *0.88*	4.46 *1.18*	4.46 *1.28*
4. I would be willing to sign a petition for an environmental cause	4.03 *1.43*	3.92 *1.05*	3.99 *1.58*	3.48 *1.20*	4.20[***] *1.58*	3.92 *1.05*	3.48[**] *1.20*	3.99 *1.58*	4.20[*] *1.58*
5r. The whole pollution issue has upset me, I feel it s not overrated (reverse scored)	3.68 *1.51*	3.65 *1.11*	3.69 *1.69*	3.47 *1.20*	3.71 *1.72*	3.65 *1.11*	3.47 *1.20*	3.69 *1.69*	3.71 *1.72*
6. If we could just get by with less, more for future generations	3.99 *1.35*	4.12 *0.96*	4.03 *1.50*	3.98 *0.98*	3.86 *1.59*	4.12 *0.96*	3.98 *0.98*	4.03 *1.50*	3.86 *1.59*
7. Manufacturers should be encouraged to use recycled materials	4.69 *0.82*	4.64 *0.63*	4.66 *0.91*	4.56 *0.54*	4.77[**] *0.94*	4.64 *0.63*	4.56 *0.54*	4.66 *0.91*	4.77[*] *0.94*
8. Future generations should be as important as current in public lands decisions	4.52 *0.97*	4.54 *0.69*	4.62 *0.95*	4.42 *0.78*	4.36 *1.32*	4.54 *0.69*	4.42 *0.78*	4.62 *0.95*	4.36[***] *1.32*
9. I would pay $5 more to use public lands for recreation	3.49 *1.60*	3.31 *1.24*	3.56[**] *1.70*	3.24 *1.10*	3.49[*] *1.98*	3.31 *1.24*	3.24 *1.10*	3.56 *1.70*	3.49 *1.98*
10. People should urge friends to limit use of scarce resources	4.14 *1.25*	4.23 *0.84*	4.09 *1.41*	3.98 *0.94*	4.19[*] *1.48*	4.23 *0.84*	3.98[*] *0.94*	4.09 *1.41*	4.19 *1.48*

continued on next page

Table H1. *Continued.*

11. I am glad there are National Forests even if I never see them	4.66 *0.91*	4.76 *0.52*	4.74 *0.82*	4.52 *0.68*	4.54 *1.37*	4.76 *0.52*	4.52** *0.68*	4.74 *0.82*	4.54** *1.37*
12. People can think public lands are valuable even if they don't go there	4.46 *1.07*	4.65 *0.59*	4.58 *0.99*	4.61 *0.58*	4.19*** *1.65*	4.65 *0.59*	4.61 *0.58*	4.58 *0.99*	4.19*** *1.65*
13. I am willing to stop buying from polluting companies	3.95 *1.35*	4.07 *0.97*	3.99 *1.38*	3.76 *0.94*	3.88 *1.82*	4.07 *0.97*	3.76** *0.94*	3.99 *1.38*	3.88 *1.82*
14. I am willing to make personal sacrifices to slow pollution	4.44 *1.05*	4.42 *0.80*	4.42 *1.10*	4.51 *0.68*	4.46 *1.36*	4.42 *0.80*	4.51 *0.68*	4.42 *1.10*	4.46 *1.36*
15. Forests have a right to exist for their own sake	4.11 *1.28*	3.93 *1.01*	4.19** *1.34*	3.92 *0.95*	4.09 *1.56*	3.93 *1.01*	3.92 *0.95*	4.19 *1.34*	4.09 *1.56*
16. Wildlife, plants, and humans have equal rights	4.28 *1.26*	4.15 *0.98*	4.34* *1.37*	3.90 *1.18*	4.31** *1.36*	4.15 *0.98*	3.90 *1.18*	4.34 *1.37*	4.31 *1.36*
17. Donating time or money to worthy causes is important to me	4.25 *1.05*	4.18 *0.82*	4.26 *1.17*	3.88 *0.83*	4.36*** *1.15*	4.18 *0.82*	3.88** *0.83*	4.26 *1.17*	4.36 *1.15*
Group Mean	**4.24**	**4.22**	**4.27**	**4.08**	**4.22**	**4.22**	**4.08**	**4.27**	**4.22**

ª Standard deviation

*, **, *** Mean difference is significant at α = 0.05, 0.01, 0.001

Table H2. Socially Responsible Management Values. Survey items are from the VOBA module of the National Survey on Recreation and the Environment.

Survey item number and statement	Full sample	East		West		Non-metro		Metro	
		Non-metro	Metro	Non-metro	Metro	East	West	East	West
18. We should actively harvest more trees for larger human population	2.88 _1.77[a]_	2.61 _1.18_	2.92** _2.00_	2.77 _1.16_	2.94 _2.14_	2.61 _1.18_	2.77 _1.16_	2.92 _2.00_	2.94 _2.14_
19. The most important role for public lands is providing jobs, income for locals	3.23 _1.53_	3.42 _1.09_	3.10*** _1.62_	3.35 _1.07_	3.34 _1.97_	3.42 _1.09_	3.35 _1.07_	3.10 _1.62_	3.34** _1.97_
20. The decision to develop resources should be made mostly on economic grounds	2.92 _1.51_	3.12 _1.10_	2.92* _1.70_	3.09 _1.06_	2.78** _1.73_	3.12 _1.10_	3.09 _1.06_	2.92 _1.70_	2.78 _1.73_
21. The main reason for maintaining resources now is to develop in future	4.00 _1.39_	4.09 _0.95_	3.90** _1.51_	3.92 _0.99_	4.10 _1.78_	4.09 _0.95_	3.92 _0.99_	3.90 _1.51_	4.10 _1.78_
22. I think public land managers are doing an adequate job of protecting natural resources	3.18 _1.31_	3.44 _0.96_	3.05*** _1.38_	3.39 _0.95_	3.24 _1.66_	3.44 _0.96_	3.39 _0.95_	3.05 _1.38_	3.24* _1.66_
23. The primary use of forests should be for products useful for humans	2.95 _1.64_	3.09 _1.19_	2.83** _1.73_	3.01 _1.16_	3.07 _2.06_	3.09 _1.19_	3.01 _1.16_	2.83 _1.73_	3.07** _2.06_
24. The Federal government should subsidize development and leasing of public lands	2.32 _1.58_	2.33 _1.15_	2.28 _1.73_	2.09 _0.98_	2.42** _1.95_	2.33 _1.15_	2.09* _0.98_	2.28 _1.73_	2.42 _1.95_
25. The government has better places to spend money than on strong conservation program	2.33 _1.48_	2.52 _1.08_	2.31** _1.58_	2.30 _1.00_	2.28 _1.89_	2.52 _1.08_	2.30* _1.00_	2.31 _1.58_	2.28 _1.89_
Group Mean	**2.98**	**3.08**	**2.91**	**2.99**	**3.02**	**3.08**	**2.99**	**2.91**	**3.02**

[a] Standard deviation

*, **, *** Mean difference is significant at $\alpha = 0.05, 0.01, 0.001$

Appendix I. Weighted Means by Demographic Breakdowns Grouped by Strategic Level Objectives

Access (see table I1).

Preservation/Conservation (see table I2).

Economic Development (see table I3).

Education (see table I4).

Natural Resource Management (see table I5).

Table I1. Access. Survey items are from the VOBA module of the National Survey on Recreation and the Environment.

| Survey item number and statement | Full sample | Objectives for the management of forests and grasslands (1=not at all important, 5=very important) | | | | | | | |
| | | East | | West | | Non-metro | | Metro | |
		Non-metro	Metro	Non-metro	Metro	East	West	East	West
1. Expand off-highway motorized access	2.41 *1.49[a]*	2.48 *1.11*	2.28 *1.55*	2.43 *0.98*	2.57 *1.97*	2.48 *1.11*	2.43 *0.98*	2.28 *1.55*	2.57** *1.97*
2. Trails for motorized vehicles	2.82 *1.47*	3.06 *1.11*	2.79* *1.60*	2.92 *0.98*	2.76 *1.75*	3.06 *1.11*	2.92 *0.98*	2.79 *1.60*	2.76 *1.75*
3. Trails for non-motorized recreation	3.75 *1.33*	3.56 *0.92*	3.80* *1.43*	3.27 *0.89*	3.82** *1.67*	3.56 *0.92*	3.27* *0.89*	3.80 *1.43*	3.82 *1.67*
4. Designate trails for specific uses	3.59 *1.43*	3.72 *0.98*	3.65 *1.63*	3.83 *1.01*	3.36* *1.63*	3.72 *0.98*	3.83 *1.01*	3.65 *1.63*	3.36* *1.63*
5. Develop more paved roads	2.62 *1.49*	2.59 *1.14*	2.57 *1.64*	2.45 *1.05*	2.76 *1.80*	2.59 *1.14*	2.45 *1.05*	2.57 *1.64*	2.76 *1.80*
6. Designate more wilderness to stop development & motorized access	3.84 *1.41*	3.70 *1.05*	3.90 *1.51*	3.28 *1.14*	3.89** *1.66*	3.70 *1.05*	3.28** *1.14*	3.90 *1.51*	3.89 *1.66*

| Survey item number and statement | Full sample | Beliefs about the role of the USDA Forest Service (1=strongly disagree, 5=strongly agree) | | | | | | | |
| | | East | | West | | Non-metro | | Metro | |
		Non-metro	Metro	Non-metro	Metro	East	West	East	West
1. Expand off-highway motorized access	2.52 *1.50*	2.67 *1.12*	2.50 *1.61*	2.70 *1.17*	2.47 *1.76*	2.67 *1.12*	2.70 *1.17*	2.50 *1.61*	2.47 *1.76*
2. Trails for motorized vehicles	2.98 *1.51*	2.95 *1.16*	2.86 *1.71*	2.84 *1.04*	3.22 *1.72*	2.95 *1.16*	2.84 *1.04*	2.86 *1.71*	3.22** *1.72*
3. Trails for non-motorized recreation	3.71 *1.37*	3.45 *1.07*	3.75** *1.49*	3.19 *0.96*	3.82** *1.59*	3.45 *1.07*	3.19 *0.96*	3.75 *1.49*	3.82 *1.59*
4. Designate trails for specific uses	3.94 *1.25*	4.11 *0.93*	3.92 *1.34*	4.00 *0.92*	3.88 *1.53*	4.11 *0.93*	4.00 *0.92*	3.92 *1.34*	3.88 *1.53*
5. Develop more paved roads	2.70 *1.57*	2.65 *1.05*	2.53 *1.67*	2.24 *0.95*	3.02** *2.05*	2.65 *1.05*	2.24** *0.95*	2.53 *1.67*	3.02*** *2.05*

continued on next page

Table I1. *Continued.*

	Full sample	East Non-metro	East Metro	West Non-metro	West Metro	Non-metro East	Non-metro West	Metro East	Metro West
6. Designate more wilderness to stop development & motorized access	3.66 *1.46*	3.58 *1.06*	3.84[*] *1.54*	3.09 *1.22*	3.54[*] *1.75*	3.58 *1.06*	3.09[**] *1.22*	3.84 *1.54*	3.54[**] *1.75*

Attitudes
about the performance of the Forest Service
(1=very unfavorable, 5=very favorable)

Survey item number and statement	Full sample	East Non-metro	East Metro	West Non-metro	West Metro	Non-metro East	Non-metro West	Metro East	Metro West
1. Expand off-highway motorized access	2.97 *1.45*	3.06 *1.03*	2.95 *1.61*	2.97 *1.00*	2.95 *1.82*	3.06 *1.03*	2.97 *1.00*	2.95 *1.61*	2.95 *1.82*
2. Trails for motorized vehicles	3.25 *1.38*	3.24 *0.94*	3.24 *1.51*	3.24 *0.96*	3.29 *1.71*	3.24 *0.94*	3.24 *0.96*	3.24 *1.51*	3.29 *1.71*
3. Trails for non-motorized recreation	3.59 *1.27*	3.54 *0.93*	3.53 *1.36*	3.34 *1.09*	3.74[*] *1.57*	3.54 *0.93*	3.34 *1.09*	3.53 *1.36*	3.74 *1.57*
4. Designate trails for specific uses	3.61 *1.33*	3.82 *0.90*	3.54[*] *1.47*	3.56 *0.90*	3.62 *1.72*	3.82 *0.90*	3.56 *0.90*	3.54 *1.47*	3.62 *1.72*
5. Develop more paved roads	3.19 *1.43*	3.21 *1.02*	3.35 *1.54*	2.89 *0.91*	3.01 *1.80*	3.21 *1.02*	2.89[*] *0.91*	3.35 *1.54*	3.01[**] *1.80*
6. Designate more wilderness to stop development & motorized access	3.45 *1.34*	3.34 *1.02*	3.35 *1.46*	3.24 *0.95*	3.65[*] *1.59*	3.34 *1.02*	3.24 *0.95*	3.35 *1.46*	3.65[**] *1.59*

[a] Standard deviation
[*], [**], [***] Mean difference is significant at = 0.05, 0.01, 0.001

Table I2. Preservation/conservation. Survey items are from the VOBA module of the National Survey on Recreation and the Environment.

Objectives
for the management of forests and grasslands
(1=not at all important, 5=very important)

Survey item number and statement	Full sample	East Non-metro	East Metro	West Non-metro	West Metro	Non-metro East	Non-metro West	Metro East	Metro West
7. Conserve and protect watersheds	4.73 *0.76[a]*	4.68 *0.58*	4.75 *0.81*	4.57 *0.72*	4.75 *0.86*	4.68 *0.58*	4.57 *0.72*	4.75 *0.81*	4.75 *0.86*
8. Preserve natural resources through policies such as no timber, no mining	4.22 *1.23*	3.99 *1.04*	4.29[**] *1.26*	3.83 *1.05*	4.29[**] *1.46*	3.99 *1.04*	3.83 *1.05*	4.29 *1.26*	4.29 *1.46*
9. Protecting ecosystems and wildlife habitat	4.58 *0.92*	4.49 *0.76*	4.61 *0.98*	4.23 *0.82*	4.62[***] *1.02*	4.49 *0.76*	4.23[**] *0.82*	4.61 *0.98*	4.62 *1.02*
10. Preserve wilderness experience	4.15 *1.28*	3.96 *1.01*	4.25[**] *1.31*	4.21 *0.77*	4.08 *1.65*	3.96 *1.01*	4.21 *0.77*	4.25 *1.31*	4.08[*] *1.65*
11. Preserve local cultural uses	3.82 *1.38*	3.78 *0.97*	3.85 *1.51*	3.69 *1.06*	3.82 *1.68*	3.78 *0.97*	3.69 *1.06*	3.85 *1.51*	3.82 *1.68*

Beliefs
about the role of the USDA Forest Service
(1=strongly disagree, 5=strongly agree)

Survey item number and statement	Full sample	East Non-metro	East Metro	West Non-metro	West Metro	Non-metro East	Non-metro West	Metro East	Metro West
7. Conserve and protect watersheds	4.61 *0.83*	4.57 *0.69*	4.67 *0.84*	4.36 *0.72*	4.57[*] *0.99*	4.57 *0.69*	4.36[*] *0.72*	4.67 *0.84*	4.57 *0.99*

continued on next page

USDA Forest Service RMRS GTR-95. 2002

	Full sample	East Non-metro	East Metro	West Non-metro	West Metro	Non-metro East	Non-metro West	Metro East	Metro West
8. Preserve natural resources through policies such as no timber, no mining	4.21 *1.27*	4.05 *1.01*	4.29 *1.30*	3.83 *1.06*	4.22[*] *1.57*	4.05 *1.01*	3.83 *1.06*	4.29 *1.30*	4.22 *1.57*
9. Protecting ecosystems and wildlife habitat	4.53 *0.98*	4.41 *0.78*	4.55 *1.08*	4.40 *0.69*	4.56 *1.14*	4.41 *0.78*	4.40 *0.69*	4.55 *1.08*	4.56 *1.14*
10. Preserve wilderness experience	4.22 *1.14*	4.22 *0.86*	4.25 *1.23*	3.83 *0.97*	4.24[**] *1.33*	4.22 *0.86*	3.83[***] *0.97*	4.25 *1.23*	4.24 *1.33*
11. Preserve local cultural uses	3.75 *1.36*	3.80 *1.03*	3.81 *1.39*	3.21 *1.05*	3.73[**] *1.78*	3.80 *1.03*	3.21[***] *1.05*	3.81 *1.39*	3.73 *1.78*

Attitudes
about the performance of the Forest Service
(1=very unfavorable, 5=very favorable)

Survey item number and statement	Full sample	East Non-metro	East Metro	West Non-metro	West Metro	Non-metro East	Non-metro West	Metro East	Metro West
7. Conserve and protect watersheds	3.91 *1.17*	3.81 *0.93*	3.91 *1.23*	3.71 *0.91*	4.00[*] *1.43*	3.81 *0.93*	3.71 *0.91*	3.91 *1.23*	4.00 *1.43*
8. Preserve natural resources through policies such as no timber, no mining	3.65 *1.31*	3.61 *1.03*	3.67 *1.38*	3.32 *0.90*	3.69 *1.63*	3.61 *1.03*	3.32 *0.90*	3.67 *1.38*	3.69 *1.63*
9. Protecting ecosystems and wildlife habitat	3.90 *1.16*	4.06 *0.80*	3.89 *1.28*	3.97 *0.83*	3.82 *1.42*	4.06 *0.80*	3.97 *0.83*	3.89 *1.28*	3.82 *1.42*
10. Preserve wilderness experience	3.88 *1.10*	3.92 *0.80*	3.87 *1.18*	3.83 *0.83*	3.90 *1.36*	3.92 *0.80*	3.83 *0.83*	3.87 *1.18*	3.90 *1.36*
11. Preserve local cultural uses	3.39 *1.37*	3.51 *0.97*	3.24[*] *1.63*	3.55 *0.95*	3.56 *1.45*	3.51 *0.97*	3.55 *0.95*	3.24 *1.63*	3.56[*] *1.45*

[a] Standard deviation
*, **, *** Mean difference is significant at = 0.05, 0.01, 0.001

Table I3. Economic development. Survey items are from the VOBA module of the National Survey on Recreation and the Environment.

Objectives
for the management of forests and grasslands
(1=not at all important, 5=very important)

Survey item number and statement	Full sample	East Non-metro	East Metro	West Non-metro	West Metro	Non-metro East	Non-metro West	Metro East	Metro West
12. Provide natural resources to dependent communities	3.60 *1.39[a]*	3.65 *1.03*	3.66 *1.53*	3.80 *0.89*	3.44 *1.70*	3.65 *1.03*	3.80 *0.89*	3.66 *1.53*	3.44[*] *1.70*
13. Restrict development of minerals	3.96 *1.42*	3.88 *1.07*	4.10[*] *1.40*	3.76 *1.05*	3.82 *1.95*	3.88 *1.07*	3.76 *1.05*	4.10 *1.40*	3.82[**] *1.95*
14. Restrict timber harvest and grazing	3.99 *1.27*	3.92 *0.98*	4.09 *1.31*	3.56 *1.05*	3.95[*] *1.58*	3.92 *0.98*	3.56[*] *1.05*	4.09 *1.31*	3.95 *1.58*
15. Make the permitting process easier for established uses	2.89 *1.55*	2.90 *1.14*	2.82 *1.69*	3.08 *1.06*	2.96 *1.91*	2.90 *1.14*	3.08 *1.06*	2.82 *1.69*	2.96 *1.91*
16. Develop a national policy for natural resource development	4.26 *1.23*	4.05 *0.97*	4.32[**] *1.31*	3.71 *0.98*	4.31[***] *1.40*	4.05 *0.97*	3.71[*] *0.98*	4.32 *1.31*	4.31 *1.40*

continued on next page

Table I3. *Continued.*

Survey item number and statement	Full sample	East Non-metro	East Metro	West Non-metro	West Metro	Non-metro East	Non-metro West	Metro East	Metro West
17. Expand commercial recreation	3.04 *1.45*	3.02 *1.00*	2.87 *1.49*	3.18 *1.03*	3.31 *1.95*	3.02 *1.00*	3.18 *1.03*	2.87 *1.49*	3.31*** *1.95*

Beliefs
about the role of the USDA Forest Service
(1=strongly disagree, 5=strongly agree)

Survey item number and statement	Full sample	East Non-metro	East Metro	West Non-metro	West Metro	Non-metro East	Non-metro West	Metro East	Metro West
12. Provide natural resources to dependent communities	3.29 *1.39*	3.26 *1.02*	3.38 *1.54*	3.43 *0.93*	3.13 *1.64*	3.26 *1.02*	3.43 *0.93*	3.38 *1.54*	3.13* *1.64*
13. Restrict development of minerals	3.95 *1.43*	3.80 *1.11*	4.06* *1.48*	3.55 *0.95*	3.91 *1.80*	3.80 *1.11*	3.55 *0.95*	4.06 *1.48*	3.91 *1.80*
14. Restrict timber harvest and grazing	3.94 *1.34*	3.71 *1.13*	4.02** *1.39*	3.09 *1.23*	4.07*** *1.47*	3.71 *1.13*	3.09*** *1.23*	4.02 *1.39*	4.07 *1.47*
15. Make the permitting process easier for established uses	2.90 *1.59*	2.94 *1.17*	2.73 *1.70*	2.94 *1.09*	3.15 *2.02*	2.94 *1.17*	2.94 *1.09*	2.73 *1.70*	3.15*** *2.02*
16. Develop a national policy for natural resource development	4.21 *1.19*	4.08 *1.05*	4.27* *1.21*	4.02 *0.81*	4.20 *1.46*	4.08 *1.05*	4.02 *0.81*	4.27 *1.21*	4.20 *1.46*
17. Expand commercial recreation	3.25 *1.53*	3.20 *1.05*	3.15 *1.63*	2.87 *1.00*	3.48** *1.97*	3.20 *1.05*	2.87* *1.00*	3.15 *1.63*	3.48** *1.97*

Attitudes
about the performance of the Forest Service
(1=very unfavorable, 5=very favorable)

Survey item number and statement	Full sample	East Non-metro	East Metro	West Non-metro	West Metro	Non-metro East	Non-metro West	Metro East	Metro West
12. Provide natural resources to dependent communities	3.45 *1.30*	3.46 *0.93*	3.57 *1.41*	3.04 *0.98*	3.35 *1.55*	3.46 *0.93*	3.04** *0.98*	3.57 *1.41*	3.35* *1.55*
13. Restrict development of minerals	3.30 *1.50*	3.57 *1.04*	3.31* *1.61*	3.13 *0.96*	3.21 *1.98*	3.57 *1.04*	3.13** *0.96*	3.31 *1.61*	3.21 *1.98*
14. Restrict timber harvest and grazing	3.50 *1.36*	3.34 *1.04*	3.55 *1.46*	2.96 *1.01*	3.61** *1.66*	3.34 *1.04*	2.96* *1.01*	3.55 *1.46*	3.61 *1.66*
15. Make the permitting process easier for established uses	3.05 *1.43*	3.16 *1.09*	3.05 *1.55*	3.24 *0.97*	2.95 *1.84*	3.16 *1.09*	3.24 *0.97*	3.05 *1.55*	2.95 *1.84*
16. Develop a national policy for natural resource development	3.51 *1.26*	3.61 *0.91*	3.53 *1.41*	3.17 *0.92*	3.50 *1.51*	3.61 *0.91*	3.17** *0.92*	3.53 *1.41*	3.50 *1.51*
17. Expand commercial recreation	3.45 *1.24*	3.31 *0.99*	3.42 *1.35*	3.34 *0.82*	3.58 *1.49*	3.31 *0.99*	3.34 *0.82*	3.42 *1.35*	3.58 *1.49*

[a] Standard deviation
*, **, *** Mean difference is significant at α = 0.05, 0.01, 0.001

Table I4. Education. Survey items are from the VOBA module of the National Survey on Recreation and the Environment.

Survey item number and statement	Full sample	East Non-metro	East Metro	West Non-metro	West Metro	Non-metro East	Non-metro West	Metro East	Metro West
Objectives for the management of forests and grasslands (1=not at all important, 5=very important)									
18. Develop volunteer programs to improve land (tree planting, etc.)	4.60 0.86[a]	4.47 0.72	4.63* 0.88	4.42 0.77	4.65 0.98	4.47 0.72	4.42 0.77	4.63 0.88	4.65 0.98
19. Develop volunteer programs to improve facilities (trails, etc.)	4.18 1.13	4.04 0.87	4.09 1.31	4.32 0.66	4.34 1.24	4.04 0.87	4.32* 0.66	4.09 1.31	4.34** 1.24
20. Inform the public about recreation concerns (safety, trail etiquette, etc.)	4.53 0.93	4.47 0.77	4.58 0.97	4.35 0.83	4.51 1.08	4.47 0.77	4.35 0.83	4.58 0.97	4.51 1.08
21. Inform the public on potential environmental impacts of uses	4.39 1.02	4.29 0.82	4.54*** 0.94	3.86 0.95	4.31** 1.37	4.29 0.82	3.86*** 0.95	4.54 0.94	4.31*** 1.37
22. Inform the public on economic value from developing natural resources	4.02 1.30	3.93 1.03	4.10 1.36	3.90 1.06	3.95 1.55	3.93 1.03	3.90 1.06	4.10 1.36	3.95 1.55
23. Encourage collaboration between groups to share information	4.15 1.20	4.23 0.80	4.23 1.23	3.93 0.85	4.03 1.64	4.23 0.80	3.93* 0.85	4.23 1.23	4.03* 1.64
Beliefs about the role of the USDA Forest Service (1=strongly disagree, 5=strongly agree)									
18. Develop volunteer programs to improve land (tree planting, etc.)	4.46 1.03	4.48 0.77	4.50 1.02	4.20 0.84	4.42 1.38	4.48 0.77	4.20* 0.84	4.50 1.02	4.42 1.38
19. Develop volunteer programs to improve facilities (trails, etc.)	4.22 1.09	4.18 0.85	4.20 1.18	4.18 0.82	4.28 1.31	4.18 0.85	4.18 0.82	4.20 1.18	4.28 1.31
20. Inform the public about recreation concerns (safety, trail etiquette, etc.)	4.50 0.93	4.50 0.71	4.55 1.00	4.42 0.72	4.45 1.12	4.50 0.71	4.42 0.72	4.55 1.00	4.45 1.12
21. Inform the public on potential environmental impacts of uses	4.48 1.00	4.32 0.78	4.48 1.10	4.46 0.64	4.54 1.18	4.32 0.78	4.46 0.64	4.48 1.10	4.54 1.18
22. Inform the public on economic value from developing natural resources	4.08 1.17	4.20 0.85	4.11 1.28	3.93 0.94	3.99 1.39	4.20 0.85	3.93* 0.94	4.11 1.28	3.99 1.39
23. Encourage collaboration between groups to share	4.15 1.17	4.21 0.82	4.12 1.30	4.07 0.84	4.19 1.42	4.21 0.82	4.07 0.84	4.12 1.30	4.19 1.42

continued on next page

Table I4. *Continued.*

Attitudes
about the performance of the Forest Service
(1=very unfavorable, 5=very favorable)

Survey item number and statement	Full sample	East		West		Non-metro		Metro	
		Non-metro	Metro	Non-metro	Metro	East	West	East	West
18. Develop volunteer programs to improve land (tree planting, etc.)	3.85 *1.25*	3.76 *0.97*	3.91 *1.26*	3.78 *0.91*	3.81 *1.67*	3.76 *0.97*	3.78 *0.91*	3.91 *1.26*	3.81 *1.67*
19. Develop volunteer programs to improve facilities (trails, etc.)	3.79 *1.21*	3.93 *0.86*	3.83 *1.28*	3.70 *0.96*	3.68 *1.53*	3.93 *0.86*	3.70 *0.96*	3.83 *1.28*	3.68 *1.53*
20. Inform the public about recreation concerns (safety, trail etiquette, etc.)	3.89 *1.31*	4.01 *0.95*	3.92 *1.36*	3.75 *0.76*	3.84 *1.73*	4.01 *0.95*	3.75* *0.76*	3.92 *1.36*	3.84 *1.73*
21. Inform the public on potential environmental impacts of uses	3.50 *1.33*	3.61 *0.97*	3.48 *1.39*	3.40 *0.90*	3.50 *1.77*	3.61 *0.97*	3.40 *0.90*	3.48 *1.39*	3.50 *1.77*
22. Inform the public on economic value from developing natural resources	3.40 *1.38*	3.38 *1.00*	3.38 *1.47*	3.42 *1.07*	3.43 *1.72*	3.38 *1.00*	3.42 *1.07*	3.38 *1.47*	3.43 *1.72*
23. Encourage collaboration between groups to share information	3.72 *1.22*	3.67 *0.90*	3.65 *1.40*	3.76 *0.80*	3.83 *1.46*	3.67 *0.90*	3.76 *0.80*	3.65 *1.40*	3.83 *1.46*

[a] Standard deviation

*, **, *** Mean difference is significant at   = 0.05, 0.01, 0.001

Table I5. Natural resource management. Survey items are from the VOBA module of the National Survey on Recreation and the Environment.

Objectives
for the management of forests and grasslands
(1=not at all important, 5=very important)

Survey item number and statement	Full sample	East		West		Non-metro		Metro	
		Non-metro	Metro	Non-metro	Metro	East	West	East	West
24. Using public advisory committees to advise on management issues	3.90 *1.20*[a]	3.83 *0.94*	3.96 *1.23*	3.66 *0.79*	3.89 *1.56*	3.83 *0.94*	3.66 *0.79*	3.96 *1.23*	3.89 *1.56*
25. Allow for diverse uses	4.07 *1.18*	4.00 *0.93*	4.16 *1.20*	4.11 *0.87*	3.95 *1.48*	4.00 *0.93*	4.11 *0.87*	4.16 *1.20*	3.95* *1.48*
26. Make management decisions at the local level (rather than national)	3.93 *1.22*	4.07 *0.85*	3.84* *1.40*	3.91 *0.84*	3.99 *1.40*	4.07 *0.85*	3.91 *0.84*	3.84 *1.40*	3.99 *1.40*
27. Increase total number of acres in the National Forest and Grassland system	3.81 *1.42*	3.64 *0.97*	3.97** *1.49*	3.19 *1.07*	3.75* *1.82*	3.64 *0.97*	3.19** *1.07*	3.97 *1.49*	3.75* *1.82*
28. Collect an entry fee to support National Forests and Grasslands	3.66 *1.36*	3.69 *0.99*	3.65 *1.51*	3.67 *1.22*	3.66 *1.52*	3.69 *0.99*	3.67 *1.22*	3.65 *1.51*	3.66 *1.52*

continued on next page

Table I5. *Continued.*

Survey item number and statement	Full sample	East Non-metro	East Metro	West Non-metro	West Metro	Non-metro East	Non-metro West	Metro East	Metro West
29. Increasing law enforcement on National Forests and Grasslands	4.01 *1.21*	3.96 *0.97*	3.98 *1.30*	3.65 *0.88*	4.12* *1.50*	3.96 *0.97*	3.65* *0.88*	3.98 *1.30*	4.12 *1.50*
30. Trade public lands for private to eliminate inholdings, acquire unique lands	3.05 *1.48*	3.07 *1.05*	2.99 *1.57*	2.96 *1.04*	3.15 *1.90*	3.07 *1.05*	2.96 *1.04*	2.99 *1.57*	3.15 *1.90*

Beliefs
about the role of the USDA Forest Service
(1=strongly disagree, 5=strongly agree)

Survey item number and statement	Full sample	East Non-metro	East Metro	West Non-metro	West Metro	Non-metro East	Non-metro West	Metro East	Metro West
24. Using public advisory committees to advise on management issues	3.84 *1.25*	4.04 *0.85*	3.74* *1.41*	3.98 *0.91*	3.92 *1.44*	4.04 *0.85*	3.98 *0.91*	3.74 *1.41*	3.92 *1.44*
25. Allow for diverse uses	4.02 *1.14*	4.03 *0.90*	4.03 *1.25*	4.17 *0.74*	3.96 *1.37*	4.03 *0.90*	4.17 *0.74*	4.03 *1.25*	3.96 *1.37*
26. Make management decisions at the local level (rather than national)	3.88 *1.34*	4.04 *0.94*	3.81* *1.44*	4.41 *0.80*	3.83** *1.69*	4.04 *0.94*	4.41** *0.80*	3.81 *1.44*	3.83 *1.69*
27. Increase total number of acres in the National Forest and Grassland system	3.95 *1.47*	3.81 *1.02*	3.90 *1.54*	3.32 *1.05*	4.15*** *1.85*	3.81 *1.02*	3.32** *1.05*	3.90 *1.54*	4.15* *1.85*
28. Collect an entry fee to support National Forests and Grasslands	3.69 *1.43*	3.64 *0.95*	3.81 *1.49*	3.51 *0.89*	3.61 *1.93*	3.64 *0.95*	3.51 *0.89*	3.81 *1.49*	3.61 *1.93*
29. Increasing law enforcement on National Forests and Grasslands	4.01 *1.26*	4.08 *0.91*	4.09 *1.31*	4.03 *0.83*	3.87 *1.68*	4.08 *0.91*	4.03 *0.83*	4.09 *1.31*	3.87* *1.68*
30. Trade public lands for private to eliminate inholdings, acquire unique lands	3.22 *1.49*	3.18 *1.17*	3.22 *1.69*	2.87 *0.88*	3.27 *1.72*	3.18 *1.17*	2.87 *0.88*	3.22 *1.69*	3.27 *1.72*

Attitudes
about the performance of the Forest Service
(1=very unfavorable, 5=very favorable)

Survey item number and statement	Full sample	East Non-metro	East Metro	West Non-metro	West Metro	Non-metro East	Non-metro West	Metro East	Metro West
24. Using public advisory committees to advise on management issues	3.36 *1.26*	3.39 *0.93*	3.35 *1.48*	3.34 *0.82*	3.35 *1.42*	3.39 *0.93*	3.34 *0.82*	3.35 *1.48*	3.35 *1.42*
25. Allow for diverse uses	3.73 *1.18*	3.72 *0.88*	3.71 *1.15*	3.76 *0.82*	3.76 *1.68*	3.72 *0.88*	3.76 *0.82*	3.71 *1.15*	3.76 *1.68*
26. Make management decisions at the local level (rather than national)	3.49 *1.32*	3.63 *0.98*	3.42 *1.39*	3.56 *0.94*	3.52 *1.68*	3.63 *0.98*	3.56 *0.94*	3.42 *1.39*	3.52 *1.68*
27. Increase total number of acres in the National Forest and Grassland system	3.52 *1.45*	3.51 *0.91*	3.42 *1.42*	3.68 *0.92*	3.62 *2.11*	3.51 *0.91*	3.68 *0.92*	3.42 *1.42*	3.62 *2.11*

continued on next page

Table I5. *Continued.*

28. Collect an entry fee to support National Forests and Grasslands	3.61 *1.36*	3.73 *0.89*	3.56 *1.55*	3.40 *1.03*	3.66 *1.63*	3.73 *0.89*	3.40[*] *1.03*	3.56 *1.55*	3.66 *1.63*
29. Increasing law enforcement on National Forests and Grasslands	3.85 *1.27*	3.85 *0.95*	3.89 *1.37*	3.48 *0.98*	3.85 *1.61*	3.85 *0.95*	3.48[*] *0.98*	3.89 *1.37*	3.85 *1.61*
30. Trade public lands for private to eliminate inholdings, acquire unique lands	3.22 *1.27*	3.27 *1.08*	3.24 *1.38*	2.96 *0.95*	3.21 *1.37*	3.27 *1.08*	2.96 *0.95*	3.24 *1.38*	3.21 *1.37*

[a] Standard deviation

*, **, *** Mean difference is significant at   = 0.05, 0.01, 0.001

Appendix J. Glossary of Terms

Attitudes   The degree to which a respondent feels the USDA Forest Service is fulfilling his or her objectives.

Beliefs   The degree to which a respondent agrees that a particular item is an appropriate role for the USDA Forest Service.

Biocentric   Having life as its center or main principle. A biocentric perspective could include any or all of the following ideas: persons and non-persons are all members of earth's community, earth's ecosystem is an interrelated totality, each organism is pursuing its own good and goals, and persons are not inherently superior to other components of ecosystems.

Conservation   Planned management of a natural resource to prevent exploitation, destruction, or neglect.

Eastern region   The following states are included in the region called "East" in this report: Alabama, Arkansas, Connecticut, Delaware, Florida, Georgia, Illinois, Indiana, Iowa, Kentucky, Louisiana, Maine, Maryland, Massachusetts, Michigan, Minnesota, Mississippi, Missouri, New Hampshire, New Jersey, New York, North Carolina, Ohio, Pennsylvania, Rhode Island, South Carolina, Tennessee, Vermont, Virginia, Washington DC, West Virginia, and Wisconsin.

Familiarity questions The following questions (included in the NSRE) were designed to gauge a respondent's familiarity with the USDA Forest Service. Each has a "True/False" format. The correct answer follows each question.

1. The Forest Service regulates hunting and fishing seasons. (False)

2. The Forest Service has Smokey Bear as its mascot. (True)

3. The Forest Service enforces the Endangered Species Act. (False)

4. The Forest Service manages national forests for recreation, timber, and water. (True)

5. The Forest Service provides visitor information and protects wildlife in National Parks. (False)

High familiarity   A respondent is said to possess a high degree of familiarity with the USDA Forest Service if he/she correctly answers 3 or more of the familiarity questions in the NSRE.

Instrumental   Serving as a means, useful. Hence, instrumental values are values based on something's usefulness.

Low familiarity	A respondent is said to possess a low degree of familiarity with the USDA Forest Service if he/she correctly answers fewer than 3 of the familiarity questions in the NSRE.
Metropolitan county	Any county considered a "central county" in a metropolitan statistical area, as defined by United States Bureau of the Census. These counties include a central city or at least 50% of the population of a central city. (A central city is a city or urbanized area of 50,000 or more.) A list of those counties that meet this criteria (as of Dec. 7 1999) was obtained from the Census Bureau website and used for all analyses in this report.
Module	A separate unit or section within the NSRE.
Non-metropolitan county	Any county which is not included in the Census Bureau's list of metropolitan counties is a non-metropolitan county.
NSRE	The National Survey on Recreation and the Environment
Objectives	Something toward which effort is directed; an aim or end of action. In the case of the VOBA, these are respondents' goals for public lands, actions, or states that they find desirable.
Preservation	To keep up and reserve for special use. In the case of natural resources, preservation would preclude any consumptive use.
Pre-test	As part of the development of a questionnaire, a draft version is presented to a sample of respondents, resembling the respondents who will complete the final version, in order to refine and improve the questionnaire.
Public Lands Values	A scale composed of 25 items concerning environmental and resource issues for public lands.
Reverse score	In order to calculate a group mean for two or more scale items which have the opposite direction, it is necessary to make the endpoints of the scale have the same meaning. To illustrate this we will use an example. Suppose we want to examine the overall preference for sweets as indicated by the preference for ice cream and pie. We have two scale items. For each a 1 indicates "strongly disagree" and a 5 indicates "strongly agree." In order to avoid the appearance of bias toward or against sweets, the two items move in opposite directions. The two items are: "I like ice cream" and "I don't like pie." Clearly a person who likes all sweets will answer 5 to the first item and 1 to the second. Conversely, someone who does not like any sweets will answer 1 to the first and 5 to the second. If these items are grouped, these two (clearly different) respondents will have the same mean for the group (3). This would give researchers very little (if any) information. In order to calculate a meaningful mean, we choose one of the items, in this example we'll choose the second, and reverse the scoring. So, an answer of 5 to "I don't like pie" would become a 1 (and we can reword the item as "I like pie"). An answer of 4 becomes 2, 3 remains the same, 2 becomes 4 and 5

becomes 1. This in effect creates a new item that corresponds in direction to "I like ice cream." Respondent one (the sweet tooth) will have a mean for the two items of 5 and respondent two's mean becomes 1. Now we have an indication of each respondent's preference for sweets. We can also calculate the mean for the whole sample for the group of items and gain information about the sample's overall preference for sweets. A similar rescoring was done for certain items in the VOBA in order to more accurately characterize overall preferences for item groups.

Scale A survey type which employs a set a statements to which a respondent provides a rating indicating the strength of his/her agreement or disagreement along a continuum. For example, a 1 might indicate strong disagreement, a 5 might indicate strong agreement, and the numbers in between might indicate less extreme preferences.

Scale item One statement of many comprising a scale.

Socially Responsible Individual Values A dimension of the Public Lands Values scale having to do with the actions of the individual related to public lands.

Socially Responsible Management Values A dimension of the Public Lands Values scale having to do with the actions of public land management agencies related to public lands.

Standard deviation The square root of the average of the squares of the deviations from the mean.

VOBA A survey of the American public's **V**alues, **O**bjectives, **B**eliefs, and **A**ttitudes regarding forests and grasslands.

Weighted mean An average of the values of a set of items to each of which is accorded a weight indicative of its frequency or relative importance.

Western region The following states are included in the region called "West" in this report: Alaska, Arizona, California, Colorado, Hawaii, Idaho, Kansas, Montana, Nebraska, Nevada, New Mexico, North Dakota, Oklahoma, Oregon, South Dakota, Texas, Utah, Washington, and Wyoming.

The Authors

Deborah J. Shields is Principal Mineral Economist, USDA Forest Service Research, Rocky Mountain Research Station, Fort Collins, CO [Tel (970) 295-5975; email dshields@fs.fed.us]; Ingrid M. Mar tin is Associate Professor, Department of Marketing, California State University, Long Beach, CA; Wade E. Mar tin is Associate Professor, Department of Economics, California State University, Long Beach, CA; Michelle A. Haefele is Post Doctoral Research Associate, Department of Economics, Colorado State University, Fort Collins, CO.

www.ingramcontent.com/pod-product-compliance
Lightning Source LLC
Chambersburg PA
CBHW081301170526
45165CB00011B/3366